# TRANSFORMING TRADITIONALLY

PERSPECTIVES ON ASIAN AND AFRICAN DEVELOPMENT  *SERIES*

*Perspectives on Asian and African Development No. 1*

# TRANSFORMING TRADITIONALLY

## Land and Labor Use in Agriculture
## in
## Asia and Africa

KUSUM NAIR

The Riverdale Company, Inc., Publishers
Maryland, U.S.A.

The Riverdale Company, Inc., Publishers
5506 Kenilworth Avenue
Riverdale, Maryland 20737
U.S.A.

*First published 1983*

© The Riverdale Company, Inc., Publishers, 1983

ISBN 0-913215-00-7

Library of Congress Number 83-61217

Printed in India by Allied Publishers Private Limited, New Delhi.

# TABLE OF CONTENTS

vi                                          *Contents*

# PREFACE

This project is the result of more than two decades of direct observation, interviews, and research in the production behavior of farmers and peasants in several countries, cultures, and cropping systems.

The two most basic factors in food and agricultural production—land and labor—are largely within the power of the individual farm household to allocate. But the differences in their usage are often striking and puzzling. The reasons are not always apparent. Why, for instance, should the average intensity and quality of land use be so poor in countries where climate permits year-round cultivation; crop yields are extremely low; labor is abundant, underemployed, and growing; poverty is widespread; and land and capital are in short supply?

According to geographer Pierre Gourou: "The same natural environment will result in different human landscapes when interpreted (and transformed) by traditional European peasant civilization, by Chinese civilization, and by modern American civilization." The fact that in most parts of the tropics "the quality of land use is very low," explains Gourou, "is primarily the result of techniques (and of the civilizations in a broader sense) and not the direct result of unfavorable physical conditions.... Human choices have been influenced much more by the level of techniques than by physical conditions."[1] But differences in the use of land (and productivity) can also be significant among societies at the same technological (and economic) level, in the cultivation of identical crops, such as paddy rice in Asia.

With regard to land use, for example, by the fourth century, A.D., cultivation was more intensive in China than anywhere in the rest of Asia, or the world. By the sixth century, A.D., Chinese peasants were practicing a "three fields in two years" rotation system in the north and double-cropping in the southern region. As for output, in spite of improved varieties, access to scientific techniques and yield-increasing inputs, rural development and extension education programs, growing markets and intense population pressures on arable land, in 1960-61 not a single country in South and Southeast Asia had yet equaled the record of China in the tenth century in exacting as much rice per unit of land, an estimated 2.3 tons per hectare.[2] But clearly, the higher yield of rice in China a millennium ago cannot be attributed to a superior technology. It was traditional. Or to the structure of rights in land. It was feudal. Or to the native fertility of the soils. As noted by Sung scholar, Ch'en Fu-liang (1137-1203), some of the best yields were obtained in the "barren" lands of Fukien and Chekiang.[3] They were due primarily to the meticulous quality of tillage of land, in which immense quantities of human labor had been invested to make land more productive—by terracing, irrigation, flood control, and drainage. Through the centuries, the most basic investment in Chinese agriculture has been investment in land. Early in the present century, when Japan overtook China's yields of paddy (and wheat), it was essentially by the same method—of intensive application of labor to improve and refine the quality of husbandry of crops and land. Later, as modern techniques, inputs, and mechanically powered equipment were adopted, labor-intensive tillage practices remained an integral feature and foundation of Japanese agriculture. Consequently, even in the early 1970s, there were 2,160 farm workers per 1,000 hectares of arable land in Japan, compared to 1,992 in China; 1,649 in Taiwan; 1,319 in Thailand; 990 in the Philippines; 852 in India; and 492 in Pakistan.[4] The number of agricultural workers per hectare is a rough indication of the intensity with which land is cultivated, and of labor inputs in the aggregate as well as in individual operations.

For policy purposes it is not necessary to trace the origin or "first causes" of established practices and attitudes toward land and work. They are mostly lost in the mist of history. But it is important to recognize that some of the most critical and enduring differences in cropping intensity and labor inputs per crop/hectare/year between countries and communities cannot be explained adequately or entirely by differences in man-land ratio;

the natural and physical environment; or economic and political organization, constraints, and incentives.

In view of the current concern for equity and social justice for the rural poor, moreover, it is also necessary to view and reassess the problems of land and labor use in the larger context of the role of agriculture in overall economic development, especially as it affects policies relating to land tenure and technology.

Thus, according to Simon Kuznets, agriculture makes a marked contribution to economic growth by (a) purchasing production inputs and equipment from other sectors at home and abroad; (b) selling part of its product to pay for purchases under (a) and also to buy consumer goods from other sectors, domestic or foreign, or to dispose of the product in any way other than consumption within the sector. The four basic reasons for increasing agricultural productivity and output are regarded as:

(i) supplying food and raw materials for urban/industrial sectors;

(ii) earning foreign exchange through exports or saving foreign exchange through import substitution;

(iii) selling for cash a "marketable surplus" to increase demand and provide a market for products of the industrial sector in the rural areas;

(iv) providing capital, and "investible surplus" to subsidize, even underwrite, the needed investment in urban/industrial sectors of the economy to facilitate or induce structural transformation of the economy considered necessary for modernization and development.[5]

In recent years, a fifth goal has been added—to improve the income, productivity, and quality of life, of the "poorest of the rural poor." Many developing countries have adopted special reform and welfare programs designed for the purpose. In practice, however, the impact of such programs has been negligible. Outside the socialist bloc of countries, a major proportion of the income accruing from technological innovations and economic growth continues to flow to the upper decile of the landowning classes. The policy is justified on grounds that income distribution determines the composition of demand. Consequently, if gains of higher production go to the starving poor, they will spend all or most of it on food—they will eat it up—thus restricting cash flows or stimulus to other sectors of the economy. This would slow down the rate of transformation to an industrial economy. The rich farmers, on the other hand, spend a substantial portion of their income on nonagricultural commodities, cars and cassettes, thus providing a market for the products of the industrial sector.

Agriculture, in other words, is viewed "as the seed stock or the investment fund for the total economy." The premise is that in concept and practice it is both possible and desirable "for the agricultural sector to make large net transfers to other sectors." Substantial amounts of net income and savings in relation to the farm output are consequently being transferred out of agriculture through trade, tax, and pricing policies, in the developing countries of Africa, Asia, and Latin America.[6] The policy raises two important questions:

(i) Why should the rural poor, who constitute the majority of the rural population in most poor countries, *not* be permitted to participate in the mainstream development process and spend their earned wages on food even if it means less surplus for the urban consumers? If there is a shortfall in supplies, foodgrains should be imported, as they have been in the past. But why starve peasants to feed the industrial workers?

(ii) Why should the agricultural sector provide the primary market and capital for urban industries rather than for its own products and development since, (a) agriculture is the source of livelihood for the majority of the total population, and (b) by every index, of wages, income, employment opportunities, amenities and services, the rural areas are less developed than the urban centers in even the poorest country?

These issues have been raised by some economists, like Michael Lipton. There has been much stronger and more widespread criticism of trade and price policies that discriminate against agriculture for distorting and depressing the needed incentives for commercial farmers to adopt "modern" inputs and equipment and thereby produce the requisite surplus for other sectors of the economy.

Widely ignored and endorsed as inevitable, necessary, and desirable, however, is the "factor contribution" or transfer of resources out of agriculture through large-scale infusion of "modern" capital-intensive technology in the predominantly peasant economies of the Third World.

The FAO *Production Year Book* (1971) listed three "Selected Indices of Scientific Agriculture"—nitrogen fertilizer, tractor, and combine harvester. Even a partial adoption of the three factors has made most of the developing economies increasingly dependent on foreign sources for technical and financial aid. Never have so many countries owed so much money with so little promise or prospect of repayment. This has increased income disparities, unemployment, and poverty in the farm sector which cannot be remedied by welfare or "basic needs" measures. But there has been no significant impact on the low level of crop yields or quality of tillage and husbandry.

Because all or most of the modern machines, spare parts, chemicals, and consultants are imported, moreover, a major portion of the farmers' income and savings is being shipped out of not just the countryside, but the country itself, across the seas to the industrial economies. This represents a loss of capital that neither the farm sector nor the national economies can afford. Most of the imports are not essential for improving land productivity or employment. And the great majority of the peasants who do not have sufficient land, access to land, or the means to purchase manufactured inputs, receive nothing in return unless it be some form of concessionary food aid necessitated by the destruction of their subsistence base, which too the developed countries must provide. *It would be clearly more rational and cost-effective to increase production by first utilizing the full potential of indigenous resources including labor to invest in and cultivate the soil as intensively and efficiently as possible.* It has been estimated, for example, that in 1980 the LDCs could have been supplied at least 80 million metric tons of organic nutrients—worth $21 billion—from organic sources compared to a projected consumption of 28.5 million metric tons of chemical fertilizers in the same countries in 1985.[7]

This book explores the implications and feasibility of adopting the above approach and strategy. It is limited to the foodgrain sector in *low-income* countries. Plantations specializing in cash or export crops are excluded. Since the study is based entirely on secondary sources, availability of data and magnitude of the problems of low productivity and rural poverty have dictated the choice of countries and farming systems.

In principle, the scope of the book is global. It is obviously too large, especially in view of the vast differences in agro-climatic conditions and cropping systems among regions and countries. The subject is highly complex and information with regard to the decision-making process of farm households is extremely limited, uneven, and fragmentary. No one really knows *why* farmers in various communities or countries allocate labor in agricultural production as they do. For definitive answers and solutions, much more research will be needed for each country and region, both to identify the key reasons for the underutilization of land and labor, and to assess the peasants' perceptions and response to suggestions for alternative patterns of work, land use, and cultural practices in order to increase their output and income. This is designed to be phase one of a two-phased project. The second phase (provided it is funded) would involve field work in some countries and regions to seek answers to some of the key questions. I hope this preliminary venture will generate wider interest and research in this approach to agricultural development and change among Third World planners and international donor agencies.

KUSUM NAIR

April 1983

# ACKNOWLEDGEMENTS

This project was funded by the United States Agency for International Development. The Land Tenure Center of the University of Wisconsin–Madison provided the institutional support and facilities. I am most grateful for their generous help and cooperation. All views, interpretations, and conclusions, however, are mine and not necessarily those of the supporting or cooperating agencies.

I have also accumulated countless debts to other institutions and individual scholars at this and other universities who generously shared their often unpublished research papers and reports, and provided the stimulus and opportunities for discussing my thoughts and ideas. I sincerely hope that the final outcome will not disappoint them.

# ASIA

# CHAPTER 1

# INITIAL CONDITIONS

1

On the eve of World War II, while China was torn by civil strife and foreign aggression, most of the rest of Asia was under British, French, Dutch, American, or Japanese control. Within five years of the conclusion of the war, China had a successful revolution, and India, Pakistan, Nepal, Sri Lanka, the Philippines, Indonesia, and the Japanese colonies in the Far East had become independent nations. Indochina joined their ranks in 1954, and Malaysia in 1957.

Thirty-five years after the war, the developing countries of Asia have the largest population, the lowest per capita gross national product, and the largest number of poor people in the world. In the three South Asian countries of India, Pakistan, and Bangladesh alone, 20 million rural households—some 89 million people—did not own or operate any land in 1980. They must depend on wage employment in agriculture for survival. Another 15 million households with a population of around 77 million people were, and still are, near-landless. They operate less than 1 acre per household, and 4 percent of the total cultivated area. They too must earn off-farm income to subsist. Together, the two groups account for about 30 percent of the rural households in the region.

Unlike in Africa and Latin America, moreover, the land frontier has virtually disappeared in Asia. There are a few scattered pockets of cultivable wasteland. But the cost of reclaiming old and new lands is very high. Future growth in food production therefore will have to be obtained by producing more grain per hectare/year from the land already under cultivation. The test of efficiency will have to be the number of people each hectare of land, rather than each farmer, can feed.

The great majority of the current unemployed and the new entrants to the workforce, moreover, will have to be absorbed in the rural sector, in agriculture, small-scale industries or handicrafts, for several decades in the future, if not indefinitely. No conceivable level of industrial development could possibly employ 80 percent of the workers in China, India, or Indonesia, for example.


2


Asia produces 90 percent of the world's rice and will probably continue to do so well into the twenty-first century. In all countries east and southeast of India and Bangladesh, moreover, rice is the staple food crop. Analysis of land and labor usage and productivity in Asia will be limited therefore to paddy cultivation and multiple cropping systems in which rice is the core crop.

In broad cultural, economic, and agricultural terms, the continent has three distinct regions—the East, Southeast, and South.

Except for the People's Republic of China, all the low-income countries of the continent, with per capita incomes of $370.00 or less in 1979, are located in South and Southeast Asia. Except for China, they employ far less labor per crop/hectare/year than the East Asian countries of Japan, Taiwan, South and North Korea. Crop yields and double-cropping indexes also are significantly lower in the south and southeastern region.

Per hectare output of rice reflects the same regional differences as in overall agricultural production. It is significantly lower in South and Southeast Asia. In 1969, however, the *Asian Agricultural Survey* concluded that if the environment sets "restrictive boundary conditions on what can and must be done to promote development, it also offers the region an opportunity of becoming the major agricultural producing area of the world. With the exception of the upper alpine reaches of Nepal, the high plateaus of Afghanistan and the temperate zone winters in Korea, the remainder of the region enjoys climatic conditions which would allow the production of food and fiber throughout the year. At present, it is only in scattered parts of the region that the intensity with which land is used for annual farm crops reaches a level of 1.4 to 1.6 crops per year. Yet experience indicates that up to four crops per year can be harvested by attentive and skilled farmers growing short duration varieties in carefully adapted rotations on land that is irrigated and drained, and kept at a high level of fertility by the use of fertilizers. Continuous cropping experiments at the International Rice Research Institute in the Philippines have produced annual yields of over 20 tons of grain per hectare...." Predicted the Survey Team: "...a good yield for a rice crop grown under sunlight in the dry season would be approximately eight tons per hectare, the same crop grown in the cloudy monsoon season could be expected to provide about six tons per hectare."[1]

The optimism was based on the release in the mid-1960s of the first tropical semi-dwarf photosensitive varieties of rice that were highly responsive to fertilizers, and therefore high yielding. Imported from Taiwan, and developed at the International Rice Research Institute (IRRI) at Los Banos in the Philippines, like the dwarf Mexican wheat varieties developed at CIMMYT (International Center for Maize and Wheat Improvement), they were described as "miracle" seeds.

In anticipation of the hope and promise, William S. Gaud, Administrator of AID, christened the phenomenon the "Green Revolution." In a symposium-style hearing in the Congressional Foreign Affairs Subcommittee on National Security Policy and Scientific Developments in December 1969, he described the new technology as a "tremendous development...also important because it added an element of drama, an element of excitement— some sex appeal if you will—to agricultural production." Not only did it make the job of increasing food production much easier, according to Mr. Gaud, but the "normally complicated business of the development process—how to get a country to develop, how to get people to change their attitudes—suddenly came down to a very simple proposition: one man seeing his neighbor doing better than he was doing." Furthermore, now it was "the South Asian peasant, the agricultural ministers of developing countries, who are demanding more fertilizer, seeking more training, asking for research institutes, better seeds, and other elements of the new agriculture. We are not pushing them anymore. They are pushing us. And that is real progress when you are talking about development."[2]

Seasoned international experts and economists predicted imminent abundance, and large surpluses of grain that would create marketing problems for the developing countries of Asia.

## CHAPTER 2

# LAND-FERTILITY AND FERTILIZERS

1

Differences in agro-climatic conditions, techniques and basic infrastructure, socioeconomic institutions and incentives, make it extremely difficult to compare, explain, or generalize about the reasons for the striking difference in rice yields in different regions and countries—*why* farmers in some areas do not use inputs at what appear to be clearly profitable levels; or *why* they are technically less efficient than appears feasible; and, alternately, *why* farmers in some other regions produce more, more efficiently, often under less favorable circumstances.

As indicated by Randolph Barker and the results of the constraints research initiated by the International Rice Agronomic Network (IRAEN) in 1974 to understand better the reasons for the poor performance of the Green Revolution technology in Asia, the maximum output possible depends on the *total* environment, including the farmers' beliefs and perceptions of costs and risks, under which the inputs are used. Thus, if "farmers do not see a particular factor as a constraint, one would not expect them to take action to overcome it."[1]

The problem is that the composition and relative importance of the most critical factors and constraints preventing farmers from applying the necessary inputs in the required manner and quantity may differ from one area to another, from farm to farm, and at different times. The measure and definition of technical efficiency in that case will be unique for each farmer.

Many of the factors, like information, prices, and supplies of manufactured inputs, moreover, are beyond the control of individual farmers or local communities to acquire, modify, or manipulate. This means that the theory that firms operate on the most technically efficient production function has little practical relevance. "The problem lies in being able to identify the cause of technical inefficiency in order to be able to suggest the appropriate solution."[2]

For policy purposes therefore it is essential to identify the behavioral constraints on the efficient use of resources in agriculture. For the limited purpose of this book, the only way to do so would be to analyze the most widely prevalent *on-farm* yield-affecting practices and use of family labor in select areas and periods of history. Practices determined primarily by the allocative choices individual farm households make could presumably also be altered if they so decide.

The more difficult problem of course is the larger, and yet *unresolved* question: *if and how millions of individual peasants can be persuaded to make similar and predictable changes in long established beliefs and perceptions of costs and risks.*

For example, according to the *World Development Report, 1978*, in South Asia, an immediate source of gains in productivity is the improvement of simple crop management practices, beginning with increased plant density and proper plant spacing, followed by the use of good seeds, seed treatment, proper tilling, weeding, and better preparation of seed beds and nurseries. These improvements are capable of raising yields substantially without any increase in inputs other than labor and better use of information.

> Recent experience in India suggests that yields could be increased through such measures by 10 to 30 percent on rain-fed land and by 25 to 50 percent on irrigated land...production increases of 1.5 to 2 percent a year in agriculture for a decade or more might be possible solely on the basis of existing infrastructure and levels of inputs......[3]

They would not involve any cash expenditure; nor any risk. They are not beyond the technical or management capacity of the average farmer in India and South Asia. Yet three decades of concerted effort have failed to induce the cultivators to make most of these simple changes in cultural practices.

2

In addition to good husbandry, perhaps the single most important requirement for higher yields in any country is good soil fertility. It is absolutely essential in both temperate and tropical climates and in irrigated and rain-fed or upland rice cultivation. That is why chemical fertilizer has been the cornerstone of the Green Revolution strategy. It was, and still is, regarded as "one of the most crucial requisites for the region, since increased food production depends heavily upon the enhancement of yields, and these, in turn, depend on improved soil fertility."[4]

Despite heavy price subsidies and fervent campaigns to promote its use, however, the average consumption of chemical fertilizers in South and Southeast Asia has remained extremely low. But again, as with all modern inputs, it is difficult to isolate the diverse exogenous factors that affect fertilizer demand across time, farm, village or country: prices, availability, and response under different conditions of soil, seasons, rainfall, quality of irrigation and crop varieties.

The levels and efficiency of fertilizer application are also affected by the farmers' perceptions of risk, expectations of return on cash expenditure incurred for its purchase, and, above all, the management practices and information which too must be acquired and refined by experience.

Thus, according to village studies in 14 locations in India, Pakistan, Malaysia, Philippines, Indonesia, and Thailand in 1971-73, the "optimum economic yields at the experiment station were consistently higher than average yields in the related village and frequently exceeded village yields by 2 t/ha." Though all the factors explaining the yield gap were not identified, most of it was associated with the difference in the level of applied nitrogen. In a number of cases, however, the yield gap was less than 2 tons per hectare. But the farmers used more fertilizer than was used at the experiment station. "Apparently, the experiment station used fertilizer more efficiently, achieving a higher yield gain per kilogram of applied nitrogen."[5]

In Bangladesh, ten years after the introduction of the high-yielding varieties, their impact on rice production was not significant. One of the consistent findings of the field trials in the rice research project area in Dacca district was that the basic constraint to higher yields was the low level of fertilizer use and efficiency.

Farmers were aware of the potential of the new seeds and associated technology. Availability of fertilizer was not a serious problem in the area in 1977. Yet "despite the proven profitability of higher levels of fertilizer" most of the farmers interviewed did not consider further increases because they believed that their level of fertilizer use was adequate. Evidently they were unaware that in most cases they could have profitably increased their yields by at least one ton per hectare with higher levels of inputs.

As noted by Ekramul Ahsan, aside from a massive infusion of capital, "successful introduction of the new rice varieties and the improved technology requires acquisition and application of new skills of husbandry and management...."[6] Hence the current emphasis on investment in human capital.

### 3

The management factor and individual variations in allocative and technical efficiency of rice cultivation should not be critical, however, in the use of traditional techniques and inputs. The process of learning is by inheritance; cultural practices are usually well adapted to the natural environment; and they are widely diffused in a common framework of goals, values, and attitudes to work and production as prescribed by custom and enforced by the community.

Soil husbandry moreover is just as important for obtaining optimal yields with traditional rice varieties as it is for the new seeds of the Green Revolution era. It may be useful therefore to narrow down the analysis of customary practices to this single aspect of rice technology in Asia, of fertilizing the soil on the individual farm.

Knowledge of the principles of and need for replenishing the land is probably as ancient as settled agriculture. The "use of manure, fallowing and green manuring all have a long history."[7] The only difference is that instead of using chemical components, the peasants used organic and natural nutrients—like making bread at home instead of buying it at the bakery or supermarket.

The use of rotations is uncommon in rice cultivation in Asia;

hence the greater need to maintain soil fertility by artificial means. Yet, curiously, differences in the level and efficiency of using organic inputs *reflect exactly the same regional patterns as in the current use of chemical fertilizers.* And, as with the latter, the differences cannot always be explained by factor prices.

Thus even prior to the prescribed linkage of the varietal revolution with chemical fertilizer in the mid-1960s when fertilizer was either not available or too expensive for use in paddy cultivation, the use of organic and natural manures was negligible throughout South and Southeast Asia. Often it amounted to little more than the bounty of fresh droppings of draft animals and the human population. By contrast farmers in the East Asian countries invested in and nurtured the land with a care and concern more akin to that of a mother for a newborn rather than of calculating economic agents.

In Japan, for instance, the number of farm animals was much lower than in China or India. And it was not customary to rest a field or rotate the crops. But throughout the Tokugawa period (1603-1867) the soil was fertilized intensively by compost made from household night soil and grass and leaves collected by the farmer from communal pastures and forest lands. One tan (0.245 acre) required 70 to 80 horse-loads of cut grass.

By the latter part of the nineteenth century, as cultivation crept up the mountain sides, sources of green manures began to dwindle. Commercial fertilizers—dried fish, Manchurian soybean cake, urban night soil, and phosphate fertilizers which Japan started producing on a commercial scale in 1888—became increasingly available. But they did not displace the use of manures produced on the farms. According to F.H. King, "in 1908 Japanese farmers prepared and applied to their fields 22,812,787 tons of compost manufactured from the wastes of cattle, horses, swine and poultry, combined with herbage, straw and other similar wastes and with soil, sod or mud from ditches and canals. The amount of this compost is sufficient to apply 1.78 tons per acre of cultivated land of the southern three main islands."[8]

In subsequent decades, there was a sharp decline in the price of purchased fertilizer relative to that of rice in Japan. In 1933-37 it was nearly one-fourth of the level in 1883-87. Yet while it was probably an important factor in accelerating an increase in the use of purchased fertilizers in pre-war years, the use of farmyard manures also increased during the same period.[a] It did not decline. There was a corresponding increase in the yield of paddy per hectare of cropped area—from an average of 2,407 kg in 1883-87, to 2,985 kg in 1903-07, to 3,657 kg in 1933-37.[10]

Both before and after the introduction of commercial and chemical fertilizers, moreover, the pattern was consistent. Even at the close of the eighteenth century, expenditure on *purchased* (organic) fertilizer to supplement the homemade manure was usually the largest item in a typical farm budget of the Japanese peasant. It accounted often for more than half of his total cash outlay compared to 45 percent in 1890 and 53 percent in 1908. In 1965, of a total annual expenditure of 150,000 yen to grow one crop of rice on two hectares of land in southern Hokkaido, farmer Yamayasu spent 50,000 yen to purchase chemical fertilizers. And he also spent the winter months carting special earth to mix with and dress the topsoil in his fields to improve its structure. In fact,

> ...as with family labor so with fertilizer, the Japanese farmer does not appear *ever* to have calculated its cost. It came to be treated as an indispensable item—part of fixed overheads. Neither price nor the marketable surplus of his produce determined the quantity of fertilizer applied. It was affected only by his financial ability to buy.[11]

As for labor, in 1888, farmers in Shonai plain in Yamagata Prefecture spent 200 hours per hectare for manuring the rice field. Even in 1956, when chemical fertilizers were widely used, the Japanese cultivator devoted 168 hours/ha to manuring paddy, compared to none in Central Luzon in the Philippines. Chemical fertilizer introduced along with the high-yielding varieties of rice in the mid-1960s was by and large the first kind of fertilizer ever used in Luzon. It did not replace manure.[12]

Similarly in Taiwan in 1939-40, only about half the rice acreage was planted to the high-yielding fertilizer responsive Ponlai variety first introduced in around 1920. Farmers then applied 10,800 kg

---

a. Contrary to the conventional view that the price of fertilizer has been the main stumbling block in the developing countries, fertilizer-rice price ratios in many South and Southeast Asian countries in the early 1960s were more favorable than in Japan in 1883-87 and in the early years of this century.[9]

(in bulk) of organic manures and 106 kg (nutrients) of synthetic fertilizer per hectare of cultivated land.

In the mid-1970s, however, only the modern varieties of rice were being grown in Taiwan. The price of fertilizer relative to rice was lower in the island than in Thailand, Indonesia, Philippines, Sri Lanka, and Bangladesh—the other five countries in which the IRAEN constraints research was conducted. And at 3.4 tons per hectare (1976), the yield of rice also was the highest in Taiwan, and particularly pertinent because of its tropical climate. It could be argued that due to favorable prices, a good infrastructure, and easy access to quality seed and inputs including institutional credit, the Taiwanese rice farmers applied much larger, probably optimal, quantities of chemical fertilizer—an average of 903 kg per hectare in 1976—than their counterparts in the other five countries. But as in Japan, although shortage of labor was a major constraint at the time, *only* in Taiwan did the farmers fertilize the rice crop with substantial quantities of organic manures as well, almost as much as in 1939-40 when labor was not a constraint.

Furthermore, the farm households made their own compost. It was not purchased. Consequently, although on certain varieties seven to ten thousand kilograms of manure per hectare was applied, the quantity used was limited by the availability of family labor and the number of livestock the family could raise. The manure was used as basal dressing at the time of initial plowing. And "most farmers with high yields appeared to use high amounts of stable manure."[13]

Since the only cost component of composting for self-employed farmers is the self-evaluated value or "implicit wage" of family labor, however, the imputed price of manure should have been significantly lower in the other five countries. If it was rational for farmers in Taiwan to make their own farmyard manure for use on their farms in the mid-1970s and earlier, moreover, it could not be less so in labor-surplus countries like Bangladesh.[a]

According to Chandler, longtime experiments revealed that from half to two-thirds of chemical nitrogen could be replaced with organic nitrogen from such sources as compost *without any*

a. The per capita income of Taiwan in 1976 was $1,070 compared to $110 in Bangladesh, $240 in Indonesia, $380 in Thailand, $200 in Sri Lanka, and $410 in the Philippines.

*reduction in yield.* "Organic sources of nitrogen increase the organic matter content of the soil, release nitrogen slowly during the entire period of crop growth, and add other plant nutrients, both major and minor." Where labor is abundant, therefore, "rice farmers can well use local sources of organic matter, rice straw and other crop residues, and animal manures, for at least half of the nitrogen required by a high-yielding rice crop."[14]

In fact, if it is at all profitable to use chemical fertilizer in any crop, anywhere, the use of organic manure should be much more profitable because it does not require cash or credit, involves no risk or dependence on world markets and domestic bureaucrats and brokers, and the raw materials are available and can be processed locally. They are free with almost zero opportunity cost, with the added advantage that the entire income from increased output accrues to the farmer. Yet through history valuable organic resources have been wasted and unwanted rice straw burnt instead of being composted and returned to the land in most countries of South and Southeast Asia.

## 4

Japan and Taiwan are small island domains with relatively homogeneous populations. After World War II they received massive infusions of foreign capital and United States aid. Both successfully implemented radical land reforms creating a unimodal system of small owner-cultivated family farms of near equal size. Neither is a developing country today. In the decades preceding the war the two countries had achieved the highest yields and rates of growth per hectare and worker in the continent. And wet rice cultivation was a relatively modern introduction from mainland China.[a] In fact, from the earliest times—in the third century, B.C.— until the seventeenth century, the Chinese tradition, techniques, and influence in agriculture generally, and rice culture in particular, were paramount throughout East Asia, including Japan.

India also radiated considerable influence on agricultural technology in South and Southeast Asia from the second century,

a. In Taiwan shifting cultivation was the more common mode of farming until its conquest by the Chinese in the seventeenth century. The efficient and intensive techniques of irrigated rice cultivation of the modern period were developed only after the Japanese annexation of the island in 1895.

A.D. onward. And though there is some controversy on the subject, India is also credited with domestication of rice—*oryza sativa.* In any case, the early maturing and relatively drought-resistant variety of rice that virtually transformed the Chinese cropping system early in this millenium was imported from the Indian state of Champa in Indochina. The Buddhist monk Shih Wen-ying gives the following account:

> Emperor Cheng-tsung (998-1022), being deeply concerned with agriculture, came to know that the Champa rice was drought-resistant and that the green lentils of India were famous for their heavy yield and large seeds. Special envoys, bringing precious things, were dispatched.... From Champa twenty *shih* (Chinese bushels) of [rice] seeds were procured, which have since been grown almost everywhere. From central India two *shih* of green lentil seeds were brought back.... When the first harvests were reaped in the autumn, [Cheng-tsung] called his intimate ministers to taste them and composed poems for Champa rice and Indian green lentils.[15]

From 1012 onward the value of the new rice began to be recognized nationally in China. It initiated the development through natural selection of many more, even earlier-maturing varieties, thus permitting not only expansion of rice cultivation to higher altitudes, but also the double and multiple cropping so characteristic now of Chinese agriculture.

> Directly by way of doubling China's rice area and indirectly by promoting a better cropping system, the long-range effect of early-ripening rice on China's food supply and population growth was prodigious.[16]

Currently, China is the world's largest producer of rice by a considerable margin; more than three times the annual output of India. But India is the second largest producer of the cereal. Rice is the single largest crop grown in both countries.

A brief but more detailed comparison of the history and cultural practices in rice cultivation in these two ancient and most populous countries in the world might be instructive therefore for evaluating various rural and agricultural development strategies and assessing the factors that influence land and labor use in food production.

## CHAPTER 3

# CHINA AND INDIA: 1950-1980

## 1

In modern times both China and the Indian subcontinent have been known as lands of abject poverty and famine. Both suffer from severe natural calamities, chronic and recurrent droughts and floods. "Floods in China imperil the lives of a greater number of people than anywhere else on earth."[1]

The land area of China is much larger—9.6 million square kilometers compared to 3.3 million square kilometers in India. But the per capita and proportion of arable land is far smaller in China—only about 10 percent of the total. The ratio of agricultural population to arable land in India is less than half as high as in China. The climate in China also is more diverse and extreme. And although most of China is in the temperate zone, the most intensively cultivated rice areas lie further south with tropical and subtropical climates influenced by the monsoon.

Both countries initiated a new era and approach to rural development through central government planning in 1950. Their socioeconomic goals were basically similar—to transform a backward agrarian society into a modern industrial nation.

Both governments aimed at mobilizing the peasants in a massive participatory and cooperative effort to generate large-scale capital construction, to intensify cultivation, increase crop production and productivity, and improve their living conditions and environment.

But whereas China embraced Marxism-Leninism and became a communist state, India opted for a liberal federal parliamentary democracy with vague socialist aspirations.

Due to the prolonged Sino-Japanese conflict in the 1930s, World War II, and the climactic civil war (1946-49), moreover, there was almost total fragmentation, demoralization, and disintegration of

the political, economic, and social fabric of China at the time of
the communist take-over in February 1949. In India, except for
disturbances and dislocations in the two border states of Punjab
and Bengal, the transition to independence from colonial rule was
smooth and orderly after a century of peace—a change essentially
of only the national status, constitution, and flag.

Subsequently also, the structure and institutions of rural India
did not undergo any basic or drastic transformation due to reform
or reorganization of agricultural production and redistribution of
income, wealth, and assets. The several redistributive reforms that
were contemplated or legislated were marginal and were either never
implemented at all, or only partially so. In China, on the other
hand, where a drastic land reform had been implemented by 1952,
90 percent of the households had been organized into cooperatives
by 1956, the cooperatives—virtually the whole agricultural
sector—had been converted into collectives by 1957, and again, the
commune system had been totally reorganized by 1961.

Nor did India witness massive policy gyrations and movements
like the Great Leap Forward of 1958 or the Great Proletarian
Cultural Revolution (1966-68) in China. Indian agriculture, in
other words, was not subjected to any radical experiments, stress,
or structural change.

The total population of China was larger than that of India—
574 million (1952) and 361 million (1951), respectively. China was
a poorer country than India. The level of illiteracy, however, was
about the same, and university enrollments a mere 0.3 percent.
There was widespread poverty, underemployment, and seasonal
idleness in the rural areas. And improper planning and poor
execution of rural development programs resulted in several, and
often serious, technical blunders in *both* countries.

The difference in ideology and political system, however, gave
India much greater access to international trade and technical
assistance. Total aid received by the country increased from $883
million in 1960/61 to $2,490 million in 1981/82. China received its
first World Bank loan of $200 million (for education) in June 1980.

In fact, after the withdrawal of Soviet technical experts and aid
in 1960 China became a virtually closed economy. During the few
years of collaboration, moreover, there was little or nothing that
the Russian scientists could teach the Chinese about agriculture. It
was generally the other way around. China "supplied the Soviet

Union with 62 items of agricultural materials (presumably research findings) and more than 2,500 varieties of seeds and seedlings."[2]

As India entered its Sixth Five Year Plan period in 1980, however, her population had increased to 659.3 million and that of China to 964.5 million (mid-1979). And although the per capita GNP in China had crept ahead of India, they were both still in the low-income category of the Third World countries with over 70 percent of the workforce in agriculture. The per capita GNP in China in 1979 was $260 compared to $190 in India. Countries with per capita incomes of $370 or less in 1979 are defined as "low income" by the World Bank.

The net acreage under food crops, however, is larger in India— 129 million hectares in 1978/79—than the *total* arable land area of China—99.50 million hectares. India also has a larger area under irrigation. It is, in fact, *the* largest in the world—56.6 million hectares (potential) and 52.6 million hectares (utilized) in 1979-80, compared to 45 million irrigated hectares in China. Of the rice acreage in India, 41.6 percent was irrigated in 1978-79, and of wheat, 65.2 percent.

In India, in 1978/79, foodgrain production was 131.9 million tons compared to 332.2 million tons in China (1979). India's greatest output of foodgrains has been 133.1 million tons in 1981/82; inadequate rainfall caused a decline to below 130 million tons in 1982/83.

The difference in total production is reflected in crop yields. Thus, the average per hectare output of just the *irrigated high-yielding* variety of paddy in India is still lower than that of China in the early seventeenth century—of 2.3 tons per hectare. In fact, the projected yield of *irrigated HYV* paddy in India for the last year of the Sixth Plan (1984-85) is only 2.231 tons per hectare, and a national average (for all rice) of 1.5 tons/ha.[3] By 1979, the average yield of paddy in China had increased to 4.25 tons/ha. Its rice production alone—143.7 million tons—exceeded the total output of foodgrains in India in the same year on more than three times the acreage.

In recent years, except for soybeans, crop yields in China have been 30-70 percent above average world levels, and considerably higher than in the developing countries. As for per hectare output of all cereals, the average per hectare output in China (2.65 tons in 1977-79) was ahead of not only the developing countries (1.46

tons), but of developed (market and centrally planned) economies as well. Between 1977 and 1980, whereas the growth of agricultural output averaged close to 7 percent a year, production of foodgrains increased by some 37 million tons in spite of poor wheat crops due to bad weather in 1977 and particularly 1980. The average annual increase in grain output at 2.5 percent over the past 25 years, moreover, is safely above the rate of population growth. It was sustained despite a decline in the cultivated acreage. And yet, as observed by Thomas Wiens,

> Chinese agriculture today is so unmechanized, so akin to gardening, that "technological development" seems a misnomer—surely there has been little change from traditional practice. But then, as one notices a hand tiller here, an electrified irrigation pump there; as one notes the application of chemical fertilizers (by hand), or remarks on the uniformly impressive condition of standing crops, one might concede that there are some elements of "modern farming" visible in the scene.[4]

2

Despite the several factors that appear to weigh in favor of the Indian cultivator, it is impossible to disaggregate or measure the contribution of individual variables, economic and technical, to the vast difference in the per hectare yields of rice (and other foodgrains) in the two countries. But again, the importance of land fertility probably transcends all other factors. And the differences in their current and historical practices and levels of fertilizing the soil on the farms are as striking as the gap in output.

Soils of alluvial origin predominate in the rice growing regions of China and India. After centuries of continuous cultivation, their fertility has been greatly eroded. They are particularly deficient in nitrogen. They also lack organic matter.

Statistics for both countries are largely approximations and calculated guesses. But they suffice to illustrate the nature and magnitude of the difference in the technology and quality of soil management and husbandry.

An analysis of the differences would be pertinent for the developing countries in not only Asia, but in other hemispheres as well, especially Africa.

In 1979-80, for example, Indian farmers applied a total of 5.26 million tons of chemical fertilizers at an average rate of only about 31.5 kg (nutrient) per hectare of cultivated land.

According to the *Sixth Five Year Plan (1980-85)*: "The amount of chemical fertilizers being applied per hectare is currently so small that in many places diminishing returns are not expected to start for a long time to come. The agronomic practices in many parts of the country," moreover, "are such that over 50 percent of the nutrients applied tend to get lost during the south-west monsoon season."[5] Usually the compost or farmyard manure is unloaded in small heaps in the open fields before it is spread. Valuable plant nutrients are lost as a result of the exposure, from drying and leaching.

China consumed a total of 12.32 million tons of chemical fertilizers (nutrient weight) at an average rate in excess of 125 kg per arable hectare in 1980. In addition, some 25 tons (in weight) or 250 kg (nutrient) per hectare of organic substances and manures were also applied. In some areas it was more than 30 tons per hectare per year. The average level of fertilization in China is the highest in the world.

Yet only three years earlier (1976-77), the use of chemical fertilizers per hectare of agricultural land in the two countries was practically identical—18.4 kg in China and 18.8 kg in India. (Agricultural land here includes land under permanent meadows and pastures in addition to arable land and land under permanent crops.) But again, in 1977 also, the Chinese farmers had applied an additional 22.99 million tons (in nutrient content) and 2,081,000,000 metric tons in gross weight of organic fertilizers— from night soil (397.97 million/t); hog manure (492.30 million/t); draft animal manure (762.72 million/t); green manure (167.6 million/t); oil cake (4.66 million/t); compost (103.60 million/t); and 152.12 million tons of river and pond mud and "other," which may include ash, leaves, weeds, and plant refuse not used as animal feed, and manure from chickens, ducks, rabbits, and other domestic animals.[6] It would have been difficult to find any cultivated land in the country on which no fertilizer of some kind had been applied. In India, on the other hand, nearly 67 percent of the cropped acreage was not fertilized in 1976-77.

Instead of being broadcast on the surface, as in India, moreover, the organic fertilizers in China are applied in layers, giving the field a number of dressings according to the dissimilar needs of different crops and at different periods of growth. In April 1979 the Fujian Soil and Fertilizer Research Institute announced the development of a pelletizer and a reportedly cheap machine that can be fabricated by agricultural plants of counties and communes for deep application of the pelletized fertilizer. It is claimed that the technique raises fertilizer efficiency by 20-30 percent and raised crop yields in test localities by 10-15 percent.[7]

Unlike China, detailed information about the sources and quantities of production and utilization of various types of organic and natural manures at the farm level is not available for India. Statistics and often even a mention of nonchemical fertilizers are generally omitted from most national and international reports, including the World Bank's "Survey of the Fertilizer Sector in India" (1979). The *Sixth Five Year Plan*, however, estimated that about 1,000 million tons of organic wastes in the form of crop residues and another 300 to 400 million tons of cattle dung and animal droppings are available annually. If recycled and utilized, they would provide an additional 6 million tons of nitrogen, 2.5 million tons of phosphate, and 4.5 million tons of potassium.

> It is estimated that the total rural compost, which could be prepared from these rural wastes, would be about 50 million tonnes; similarly urban wastes could also contribute about 15 million tonnes of compost.[8]

And once again, the *intent* of using "on a large scale the technology of composting based on locally available materials, including the designing of simple and low cost composter for rural areas as well as economical big composter for urban areas," was reiterated by the Indian government and planners for the sixth time in thirty years. As observed by the Working Group on Energy Policy report,

> The level of use of agricultural waste has not been satisfactorily estimated so far.... From time to time, some schemes for the use of animal waste and agricultural waste in certain urban areas have been mooted but no systematic effort for decentralised agricultural waste utilisation has been made seriously.[9]

The report also pointed out that the technology of biogas production from animal dung on an individual family scale had been fairly well developed and the strategy was "to encourage mainly individual effort with the result that the programme has benefited individual families of a certain category in rural areas." While it has "served the need of bringing the technology to the rural areas," however, "it is not adequate to harness the sizeable biogas potential and achieve the objective of the benefits reaching a larger section of the rural community."[10]

3

Use of chemical fertilizers in food crops is of comparatively recent date in both China and India—mainly a postwar phenomenon. Traditionally peasants in both countries used various types, quantity, and quality of organic nutrients. In 1950, however, whereas animal excreta and household night soil comprised 78 percent of the total organic fertilizer used in China, over 50 percent of fresh cattle dung, the main source of manure in India, was used as fuel. Indian farmers never composted night soil due to custom and cultural prejudice.

The persistent failure in India to convert organic and natural wastes into soil nutrients, moreover, is as much a reflection of government policies and attitudes since 1950 as of the cultural legacy of the cultivating households. It also underlines a fundamental difference in the official philosophy, policies, and approach to modern technology and the use of human labor between China and India.

Thus, depletion of soil fertility "due to persistent neglect of land" was recognized as a major problem in India in the *First Five Year Plan (1951-56)*. But although increased utilization of "farmyard manure and oilcakes, bonemeal, etc." was recommended, equal or greater emphasis was placed on promotion of chemical fertilizers. The planners did not consider it necessary first to mobilize fully all the manurial resources of the organic type because "the process is bound to take some time as it necessitates the disturbance of age-old habits.... The two processes should and can go on simultaneously. Both these types of manure are necessary for maintaining and increasing soil fertility."[11]

Actually, the two processes did *not* go forward "simultaneously."

As conceded by the *Review of the First Five Year Plan*, "on the whole agriculture departments and extension agencies did not devote sufficient attention to the development of local manurial resources."[12] By 1960, the "modern tendency" was to move away from mobilizing farmers' own labor to fertilize the soil and instead provide "specialized services" for unpleasant jobs of a routine nature.

> While the Agricultural Departments propagate the usefulness of preparing the compost properly and carefully, the social education in the country seems to be working in the opposite direction. Cowdung is increasingly being considered as something messy and insanitary. The possibility of developing professional compost makers, who may receive the dung on payment and set up compost factories or cowdung gas plants needs to be explored....[13]

By the end of the Second Five Year Plan (1956-61), in fact, chemical fertilizers had overtaken farmyard manure in importance and priority. But the increase in the use of chemical fertilizers also was very slow because of high price, inadequate supplies, and lack of fertilizer responsive crop varieties and responsive farmers.

Chinese planners, on the other hand, did not pay any attention to chemical fertilizers in the 1950s. The emphasis was almost exclusively on further intensifying the utilization of traditional sources of natural fertilizers which had already been very intense for a long time.

The communist government's pressure to utilize natural fertilizers more intensively was so strong that farmers in Kiangsu province, for example, collected so much river and pond mud that, according to the *Bulletin of Eastern China Agricultural Science*, "virtually all rivers, ponds, and aqueducts in the province were 'bottom up' by 1957."[14] Nationally, collection of river and pond mud increased by one-third—from 114.09 million tons in 1952 to 152.12 million tons in 1957. Furthermore, whereas chemical and cake fertilizers (and sometimes urban night soil) were manufactured and distributed by the government and involved cash costs, most of the manures were processed locally, requiring mainly the farmers' labor.

Only in the mid-1960s did the Chinese government begin to

promote chemical fertilizers. But as indicated earlier, this did not reduce the importance of organic manures. Their application more than doubled between 1952 and 1977. The planted acreage to green manures alone increased from 3.42 million hectares in 1957 to 12.3 million hectares in 1977, despite the acute pressure on a diminishing land base.

In India, by contrast, the advent of the high-yielding varieties of rice and wheat in the mid-1960s heralded a new high-pay-off input strategy and the virtual demise of programs to promote the use of farmyard compost. Undiluted application of science, technology, and "modern" inputs now became the "key-note" of agricultural policy. It was based on the premise and dogma that "after centuries of cultivation our forefathers had already reached the optimum in the use of traditional methods.... traditional seeds and manures—they do not give a dramatic increase in yields."[15] An Indian commentator expressed the prevailing attitudes and beliefs of officials and intellectuals as follows:

> The real divide in India today is between traditionalism and modernity, revivalism and progress, conservators and innovators, the superstitious and the scientific and rational.[16]

There was no ambiguity about the choice. Chemical fertilizer is "modern," and manure is "traditional." Increase in fertilizer consumption became therefore the consuming interest and objective of government policy. And it remained so despite the subsequent steep rise in the unit and total price of imported fertilizers and the burden of subsidies provided to make their use in food crops economic and attractive.

Thus the cost of imported fertilizer increased from US $26 million in 1950/51 to $923 million in 1980/81. Despite the low level of application it is the major purchased input in crop production in India today. Subsidies on imported and domestically produced fertilizer cost the government $600 million in 1978/79.

The oil crisis and worldwide shortage of chemical fertilizers in the early 1970s restored a measure of respectability to "manure" in the planning councils of New Delhi. The government rediscovered the vast unutilized potential of local manurial resources and reaffirmed its resolve to expand rural composting programs. The Union Minister for Agriculture wrote to the State Chief Ministers on 2nd of May, 1973, to organize a massive seven-day campaign in July for popularizing compost for agricultural production.[17] Yet in the same year, 1.3 million tons of oil cakes, which have a high nutrient content and absorption rate, were exported out of the rural areas and the country. Over the next four years (1975-79) the value of exports of oilseed cakes, meals, and rice bran, averaged over $200 million per year—about 14 percent of all agricultural exports. Chinese farmers used 4.66 million tons of oil cakes to fertilize the soil in 1977. The quantity was limited mainly by the scarcity of land since oilseed crops compete with other crops.

In tune with the spirit of "modernism" moreover, despite severe shortages and high price of fuel, and an equally severe problem of idle labor, an official note entitled "Efficient Use of Inputs" suggested in 1973 that use of silt be popularized by providing small-sized bulldozers to village *panchayats* "when there are many tanks with accumulation of silt. This arrangement will augment the water storage capacities of the tanks besides making large quantities of silt available for manures."[18]

Farmers' organizations were similarly advised to mount a big campaign for control of weeds—since they "eat away 30-40% of plant nutrients applied to the soil...by *mechanical* means as well as with the application of *weedicide* which is already available in the country."[19] (Emphasis added.)

More importantly perhaps, the objectives of the recycling policy became so scientific and all-embracing—not just to help the utilization of natural manures but also to improve "the environment of human habitat" and the "ecosystem"—that organic wastes now had to be processed by "modern" methods in "modern" plants, and not in compost pits on the individual farm.

Increasingly, therefore, the government has assumed the responsibility to *produce and supply* to the farmers "quality" compost "enriched" with plant nutrients extracted from materials, such as "crop residues, tree wastes, weeds, urban and rural wastes, animal wastes including the dung, litter, droppings and carcasses, marine landings and sea weeds...."[20]

Thus, manure has been elevated to the status of chemical fertilizer. If the program is successful, not only would the farmer not be expected to dirty his hands and use primitive methods to make his own compost, but there may be no waste materials available for him to make it with (or to burn) even if he wanted to.

4

Underlying the zeal in India for *modern* methods is another basic philosophy and concern, namely:

> Reducing the drudgery of labour, while raising productivity, must be regarded as an important component of employment and basic needs policies and programmes in the agricultural sector.[21]

It is reinforced by a deep and genuine conviction among intellectuals and economists that "barring the farming techniques recently adopted by a few rich farmers, the pattern of production is already rather excessively labour-intensive, almost amounting to drudgery."[22]

According to Webster's dictionary the word "drudgery" means hard, dull, uninspiring, monotonous or menial work. But then, except for people who grow plants (usually in their backyard) as a hobby, all farm work is hard, dull, and monotonous, even in a modern mechanized agriculture, as any hog or dairy farmer in America will testify.

A self-employed farmer in Kansas combining 1,000 acres of wheat will describe the day's work as awfully "itchy" but certainly not inspiring. And at peak seasons, of sowing and harvesting, he will put in more work hours in a day than the Indian farmer does because his equipment has lights. He can therefore, and usually does, continue to work after dark when all work in the fields ceases in India.

Similarly, the overwhelming majority of the industrial wage earners in modern manufacturing plants spend a lifetime doing the same repetitive, dull, and often highly strenuous or stressful work, 6 to 8 hours a day.

Whether a particular kind of work is "menial" depends however on the cultural milieu and values of the community. To the "Anglos" in Australia, for example, stoop work is menial. But not so to the Italian immigrants, who therefore monopolize vegetable production in the Province of New South Wales. In Vermont (USA) even idle workers will not pick apples in their own state.

There is no real justification therefore for characterizing farm work in India as "drudgery." And there is even less justification or truth to the claim that the pattern of production or land use is "excessively labor-intensive."

Nevertheless, the belief that it is so has been a prime reason for the conspicuous reluctance in India to emulate the Far Eastern model of intensive small-scale farming. A committee appointed by the ruling party soon after independence to make recommendations about agrarian reforms, for instance, acknowledged the "remarkably high gross yields per acre," but rejected the system because "peasant farming in China as well as in Japan is characterized by heavy physical labour of small farmers."[23]

Preparation of organic manures of course requires enormous amounts of labor. The difference in fertilizing practices is reflected in the amount of labor utilized for manuring the rice paddies in China and India. Nationwide statistics are not available for the two countries. But Buck's survey of 1929-33 clearly indicates that even when tenancy cultivation was the rule and compulsions of a centralized socialist state were absent, the number of workdays spent on manuring was much greater in China than in India, then or now.

Thanjavur, for instance, is one of the richest and more intensively cultivated rice-growing districts in India, with a cropping intensity of 150. In 1967-69, over 90 percent of the acreage was irrigated. The total input of human labor for manuring ranged from 6.1 to a maximum of 12.1 man-days per hectare per crop depending on the season and variety of paddy cultivated.[24]

In the 1970s the peasants in Hai-ch'eng county of Liaoning province spent 18.2-23.7 man-days per year for processing and application of manure accumulated from *each* pig. In 1979 there were 320 million pigs in China.[25] Rawski estimates that the *national* average labor input for fertilizing the soil in China (1975) was around 130 man-days per hectare of sown area. It would have been much higher in the irrigated areas.[26]

Maoist policy has been severely criticized by many economists for *not* modernizing agricultural technology between 1952 and 1965; for preferring to generate agricultural growth instead by mobilizing "cheap" non-industrial resources, and using labor too intensively, "literally as free goods with no opportunity cost." As observed by Kang Chao, Chinese agriculture had been pushed close to the limit along the traditional production function and "the pay-off for further intensification of cultivation was already in a state of diminishing returns." The effort to utilize traditional inputs more intensively during the 1950s had only limited success, according to him, because traditional inputs are "neither inexhaustible nor truly costless."[27]

Nevertheless, considering the fact that resource mobilization was a key goal of Maoist policy—as it was in India in the 1950s—the massive injection of additional inputs succeeded in increasing the aggregate agricultural input index in China at a rate "probably unmatched by other nations, either historically or contemporarily. Japan's index, for example, rose by about 0.5 percent per year (compound rate) over a sixty-year span from 1880 to 1940. In contrast, the annual rate of increase for Communist China stood at 4 percent during 1952-65."[28]

At that stage of development of the Chinese economy, how else could it have been accomplished except by using traditional inputs including labor more intensively?

CHAPTER 4

# HISTORY AND PERSISTENCE OF TRADITION

1

Artificial fertilisers are used as little in China
as they are in India; but there is no organic
refuse of any kind in that country [China]
which does not find its way back to the fields
as fertiliser.[1]

1926.

Most farmers collect cattle-dung from their
cattle-sheds, but they allow the urine to go to
waste. It is rich in plant foods. As such it
should not be wasted.[2]

1973.

...no litter is supplied to the cattle, and not
once in a thousand times is any attempt
made to save the urine.[3]

1891.

The above statements underline the remarkable continuity and
persistence in both countries of their peasants' attitudes, practices,
and allocation of time and effort to soil husbandry through a
century of revolutionary changes in the social, demographic,
economic, and political environment and institutions.

John Augustus Voelcker, Consulting Chemist to the Royal
Agricultural Society of England, made the first comprehensive
survey of manurial practices in India in 1889-91. According to his
Report,

...the Indian cultivator does not make full use of what he has at hand. These are, firstly, the non-utilisation of night-soil; secondly, the imperfect conservation of the ordinary manure from cattle.... The solid excrements are picked up, and either made at once into cakes for burning, or else they are thrown on the manure heap, such as it is. The urine sinks into the ground.... Now and again a little of the softened earth is scraped away and thrown on the manure heap, but it results in little more than a deeper hollow being made...."

The loss was further compounded by the fact that "the manure is often put, not in pits, but in loose heaps into which sun and rain can easily penetrate. Even when pits occur, they are often not much more than holes dug in the ground." The instances of manure being properly preserved were "very rare."[4]

A quarter century later, the Royal Commission on Agriculture found that there had been little "advance in regard to the conservation of manure since Dr. Voelcker wrote his report.... The practice of providing litter for cattle is rarely, if ever, adopted.... No efforts are made by the cultivator to preserve cattle urine. Manure pits are still seldom found in Indian villages. Where they do exist, no attempts are made to preserve the manurial value of the contents...."[5]

Between 1872 and 1921, there were 62 million more people in British India alone. There was some increase in total production of foodgrains due to expansion of acreage under irrigation and cultivation. But according to the Commission, "it is doubtful if any appreciable increase in yield can be attributed to the adoption of better methods of cultivation or the increased use of manures." India had more animal units per cropped acre than China at that time, and possibly a slightly larger human population as well in the rural areas.[a]

The Commission also concluded that an overwhelming proportion of the agricultural soils of India had probably reached a state "of maximum impoverishment many years ago." Land productivity had stabilized, depending "almost exclusively on the recuperative effects of natural processes in the soil to restore the

a. According to the 1931 census the rural population in India was 300,700,000. According to Buck's estimate, based on his own survey and data from other official sources, the total farm population in the eight agricultural regions of China was 300,190,000.

combined nitrogen annually removed in the crops, for but little of this is returned to the soil in any other way." In China at the time, the Commission pointed out that not only "is all human waste carefully collected and utilised, but enormous quantities of compost are manufactured from the waste of cattle, horses, swine and poultry, combined with herbage, straw, and other similar waste. Garbage and sewage are both used as manure."[6]

Earlier, King had described the practices more graphically:

> [a]lmost every foot of land is made to contribute material for food, fuel or fabric. Everything which can be made edible serves as food for man or domestic animals. Whatever cannot be eaten or worn is used for fuel. The wastes of the body, of fuel and of fabric are taken to the field; before doing so they are housed against waste from weather, intelligently compounded and patiently worked... to bring them into the most efficient form to serve as manure for the soil.[7]

According to John Lossing Buck's survey of 15,316 farms in 152 localities and 22 provinces in 1929-33, of which 83 percent were in the rice regions, 90 percent of the farmers fertilized the crops. In Szechwan and southwestern regions, the percentages rose to 98 and 99, respectively. The largest amount of manures per crop acre—22,195 pounds—was moreover produced in the southwestern rice region where an Indian-type of monsoon climate prevails. Manure applied to rice was 50 percent higher than for other crops.[8]

2

> The cultivator fully realizes the value of farmyard manure for the purpose of increasing the yield of his crops. The problem does not, therefore, consist in convincing him that cattle dung is better used as manure than as fuel, but in providing him with an alternative fuel, as cheap and as useful as cattle dung.[9]

1945.

This has been one of the most basic assumptions underlying the policies to solve problems of soil fertility through the century in India.

Yet there are other countries that were as or even more densely populated than India by the turn of the century. Their farm families also needed fuel for cooking. Since the belief that livestock dung in India is burned *only* out of necessity is so firmly established, it may be useful to investigate in some detail how farmers in China have managed to cook their meals without it.

Fuel has been a more serious problem in China for a longer period perhaps than in India.

> The harm of over-population is that people are forced to plant cereals on mountain tops and to reclaim sandbanks and islets. All the ancient forestry of Szechwan has been cut down and the virgin timberland of the aboriginal regions turned into farmland. Yet there is still not enough for everybody. This proves that the resources of Heaven and Earth are exhausted.[10]

So wrote Wang Shih-ho in 1855-56. According to Ping-ti Ho, the rice culture in China proper had probably reached its saturation point by about 1850. Yet nowhere, at no time, apparently did the Chinese farmer burn animal dung. Use of coal was severely limited. And by 1930, China had less than 9 percent of its area under forest compared to 13.1 percent in British India. According to Buck, the fuel problem was so severe that most families took pains to limit the use of all fuel. "In Yenshan Hsien, Chihli, fuel is so expensive that farmers do not heat water for tea but drink it cold. On the farms in Kiangning Hsien (T), Kiangsu, which are two to five kilometers outside the Taiping gate of Nanking City, grass for fuel is a good money crop and the farmers in order to save fuel cook their rice only once a day...."[11]

The type of fuel used varied with climate and topography in various regions of China. For example, in the great plains regions of North China, where there is little hilly land for growing grasses or bushes, most of the fuel consisted of stalks or straw, the by-products of crops, compared with mainly dried grasses and bushes in East Central China. In Wusiang Hsien, Shansi, a mountainous region, with the exception of a little firewood, it was grass and bushes produced on the mountains. The firewood was taken from trees grown on the farmstead or along the edges of the fields. In Kiangning Hsien (T), Kiangsu, also, farmers produced firewood for household use largely from trees grown around the homestead. In the six provinces surveyed by Buck, the farm furnished 88.7 percent of the fuel and light consumed by the farm households. When firewood was used,

> ... sticks of wood only partially burned at the time food has finished cooking are taken out of the stove and are saved for another time by pouring water over them.... Oil for light constitutes only a small proportion of the fuel and light expense. Few farmers can read or write and many farmers are so poor that they retire early in order to save the expense of oil.[12]

3

What is put under the kettle is worth more
than what is inside.

An old Chinese saying.

In India there is a seeming reversal of priorities. It is often explained by the cultural preference for dung as fuel, especially for boiling milk, because "it gives a slow fire which does not need any attention, whereas a woodfire does. There are also ideas," as Voelcker discovered, "that cow-dung imparts to the food a particular flavour which the people like...."[13]

It would explain why Indian immigrant laborers in Burma persisted in burning cow-dung for cooking although an abundant supply of firewood was readily available. But then, if the preference for dung as fuel is so strong in India, why do the farmers rely almost exclusively on cattle manure for fertilizing the soil? Why have they ignored various other rich sources of composting materials, such as night soil; droppings of other domestic animals, birds, and poultry; house and street refuse; grass, leaves, weeds; crop wastes and residues; fish and bone meal; oil cakes; and silt from lakes and river beds? Historically, these have either not been utilized at all or only to a limited extent in a few areas by certain caste groups growing specialized crops.

"Practically, therefore, everything centres in cattle-manure, and the question of how to use it to better advantage."[14] It is the only universal fertilizer.

Then again, if the farmers elected to rely on a single and increasingly scarce resource for manure, why have they never cared to provide litter for their cattle to save the nutrition-rich urine? As noted by Voelcker, leaves were collected for parching grain, as they still are, but "neglected for litter."[15] In fact, why has the Indian farmer never made good quality manure from the dung that he does not burn for fuel?[a]

The nonuse of night soil for composting in India is similarly attributed to insurmountable cultural constraints like caste taboos, prejudice, and aversion to handling of human waste. But again, if the farmers understood its value as manure and could not overcome the cultural constraints, why did they not have it composted by their low caste farm laborers who are "born" to perform precisely such "menial," "dirty," and "polluting" tasks and are willing to do them for a pittance?

The 1973 circular to State Chief Ministers regarding the observance of the "compost week," for instance, recommended trench latrines for composting the night soil in the rural areas. And it suggested that a part-time "sweeper" be employed to spread dry refuse or earth over the night soil to ensure proper decomposition and sanitary conditions.

Historically, in fact, village servants, serfs, or bonded labor were an integral part of the structure and organization of agricultural production and labor supply and use. Their status and obligations were mandatory, hereditary, and generally attached to a particular piece of farmland and its owner. The landowner was obliged to employ them, provide a free residence site, and pay customary wages in kind. They sufficed to keep the workers alive, but not much more.

In 1801, Francis Buchanan recorded that in South Malabar (now in Kerala) a farmer owning 35 acres required "5 ploughs and 10 oxen, and 5 families of slaves." He hired a servant "to superintend his slaves." According to the 1891 census, in the rice-

---

a. In the 1960s, the average nutrient content of bulky farmyard manure in India was 0.75 percent compared to 2.3 percent in European countries. In the latter part of the nineteenth century, in Britain also cattle were the main source of farmyard manure. But the quality of the manure was far superior to that of India.

growing province of Madras, about 30 percent of male agricultural
laborers were attached laborers. In certain areas there were hardly
any agricultural laborers apart from the "slaves."[16]

Members of the laborers' households provided various services
to their masters in the home and the field. The lowest in the scale
of outcastes cleaned the privies, cesspools, and drains, and swept
the street in front of the master's house.[17]

In many areas they also kept hogs and earned a small income by
selling the bristles. Sometimes they also sold fuel cakes made from
dung which they collected from the grazing grounds and roadsides.
But they did *not* make or sell hog manure, presumably because of
lack of demand. In China in 1977, hogs supplied 492.30 million
tons of manure.

According to the 1931 census for all India, there were 407 "farm
servants plus field laborers" to every 1,000 "ordinary cultivators."
After independence bonded labor was merged into the broader
category of "scheduled castes." The current population of
scheduled castes in the country is about 100 million, of whom 52
percent are agricultural laborers. Generally landless, they continue
to retain their caste identity, status, and occupations. The system
of bonded labor was abolished by law in 1975-76. But this did not
mean an end to the institution.[18]

Curiously, the behavior of the few farm communities which
traditionally did *not* burn dung for fuel in India also remained the
same and indifferent to environmental changes. Thus Voelcker too
was convinced that although the unfortunate practice of burning
dung was "a general one," it was "rather from necessity than from
want of knowledge of its worth.... the reason why they burn dung
is that they have no wood." In support of his argument Voelcker
cites the example of a small area in Gujarat. Manures were widely
used. And presumably because the countryside was well wooded,
"no *Charotar Kunbi* (the best cultivating caste) burns dung, not
even for cooking purposes."[19] By 1959, however, the woods had
disappeared in that same area due to population pressure and
extension of cultivation.[a] Yet,

      ... every field has a raised boundary which serves to conserve

a. The district has the highest density of population in Gujarat. In the 1950s,
moreover, some 53 percent of the farm holdings were less than 5 acres and another
28.2 percent were between 5 and 15 acres.

moisture, and on which grows a cactus hedge to keep out stray cattle. Along the hedge trees are grown to provide wood both for construction purposes and for fuel, so that cow-dung is not burnt but saved for making manure...[20]

The woods had gone. But the tradition of *not* burning cow-dung for cooking had not changed. Instead the farmers had found new ways of growing their own fuel requirements.

## CHAPTER 5

# EVALUATION OF POLICIES AND PERFORMANCE

"Labor first;
Capital will follow."

A Chinese adage.

It is entirely possible that under its "modernization" programs, chemical fertilizers may some day totally displace farmyard manures in China.

> Chinese agriculturists uniformly prefer green manuring and organic recycling over the use of chemical nitrogen fertilizer, but are becoming increasingly aware, especially under the present leadership, of less unfavorable cost-benefit ratios, both in terms of labor and food production.... Certain practices with labor and capital related high cost-benefit ratios, such as the recycling of river silt and sludge, biogas production, and the cultivation of certain green manure crops, may soon be discarded.[1]

Even if that were to happen, however, Chinese policy to delay the switch to chemical fertilizer and other manufactured inputs and equipment has important advantages over that of India and other low-income countries whose governments have generally sought to supplant rather than supplement traditional techniques with the latest and the "best" technology in use in the Western world.

Thus, briefly, the initial and continuing emphasis on the use of organic nutrients in China helped improve the texture and productivity of the soil in far greater measure than in countries that have relied exclusively on chemical fertilizers.

It is essential to understand that while inorganic substitutes exist for organic sources of nitrogen, potassium, and phosphorus, there is no substitute for organic matter itself. The proteins, cellulose, and lignins that comprise plant residues increase the porosity and water-holding capacity of soils, thereby serving to prevent erosion and provide the conditions needed for good root development. Organic matter is also the food for the soil's microbiological life, which slowly converts and releases nutrients (and trace elements) in forms and at rates that plants can assimilate.[2]

The Chinese peasants, moreover, were familiar with the traditional manuring techniques. They could be further improved and refined therefore without the prior acquisition of sophisticated new knowledge and skills needed to use chemical fertilizer (and other modern inputs) efficiently—*skills which no government can possibly teach to millions of illiterate peasants in the time it takes to deliver the fertilizer to their fields*.

This is one of those rather simple and obvious facts (not opinion) which has been overlooked by the planners and policy-makers in most Third World countries.

In 1949-50, around 80 percent of the population in China was illiterate compared to 72 percent in India; 81 percent in Pakistan; and 82 percent in Bangladesh in 1970. There was an acute shortage of trained personnel including accountants. In the 1950s, many agricultural cooperatives had bookkeepers who could neither read nor write.

The number of extension service stations increased from 10 in 1950 to 13,669 in 1957. To meet their staff needs of 95,000, the stations were forced to hire large numbers of persons who had been "laid off by other organizations but had no knowledge about agriculture" or who "were under fourteen or fifteen and did not have adequate common sense." According to a 1957 survey only 5 percent of the extension station employees were college graduates, and 63.3 percent had only elementary school education. The survey noted:

In view of the low quality of cadres, those stations could make little contribution to the improvement of farm technology in our country. As a matter of fact, those cadres spent most of their time assisting local governments in administrative work.[3]

But unlike their counterparts in most other low-income countries entrusted with complex know-how about complex techniques which neither they nor the peasants were qualified to comprehend, the Chinese extension agents at least could not do serious damage—like burning crops by recommending untested and lethal doses of chemical fertilizers.

Aside from a severe shortage of technical personnel, China also had, and in many respects still has, a grossly inadequate organization and infrastructure of storage, roads, and transportation facilities in the rural areas.[a,4]

Decentralization of production of organic, and later, of chemical fertilizers, at the farm and village level, ensured supplies and timely delivery at low cost of a critical input throughout the country. Even in 1977, nearly two-thirds of nitrogenous fertilizers was produced in more than 1,400 small plants in the communes with a production capacity of 3,000 to 5,000 tons of liquid ammonia each per year.[b]

Most importantly, perhaps, the policy and emphasis on prior utilization to the maximum extent possible of indigenous materials and equipment, saved the country heavy capital investments and foreign exchange for importing modern fertilizer plants and the finished products at a time when both capital and foreign exchange were scarce. This avoided the burden of subsidies on the price of fertilizer. And it did not extract capital out of the rural areas, the cash that farmers would have paid for its purchase.

In the first decade (1952-61), for instance, the Chinese farmers saved the cash equivalent of the price of 121.8 million (nutrient) tons of locally produced organic manures from waste materials by their own labor compared to the cost of 2.024 million tons of chemical nutrients that were imported in the same period. A similar saving was made in the area of plant diseases and insect pest control. In 1958, 80 percent of the pesticides used were native

a. In 1980, China had only 890,000 kilometers of roads, most of which could not carry heavy trucks or buses and no traffic at all in bad weather. The shortage of roads was aggravated by a very limited supply of vehicles in the rural areas.

b. The first factory to manufacture ammonium sulphate established by the Indian government in 1951 had an installed capacity of 90,000 tons a year.[5]

drugs made from plants and minerals available locally. In 1976, indigenously developed biological methods of plant protection were being used on 3.47 million hectares of land. As described by Harold Reynolds:

> Some 1,500 people in one commune visited are involved in plant protection work. Of these, 65 raise and herd some 220,000 ducks—a traditional method of controlling paddy insects.... The country has the manpower to grow and deploy [wasps, fungi, bacteria, and other biological control agents] by methods economically infeasible here.... Masses of 300,000 wasps per acre are released....[6]

The methods may be economically infeasible in America, but not in labor-surplus countries like India or Bangladesh where crop losses continue to be heavy in spite of increasing use of chemical pesticides at increasing cost to the administration and the farmer.

Despite its isolation and shortage of scientists and modern research facilities, in fact China made notable advances in other areas as well. It developed its own high-yielding varieties of wheat, rice, a rice hybrid (the first in the world), and hybrid corn and sorghum. The first improved semi-dwarf rice was released seven years before the release of the IR8 variety by the International Rice Research Institute which inaugurated the Green Revolution in South and Southeast Asia.[7]

Finally, by not disdaining manual labor and traditional techniques, in less than two decades China succeeded in achieving a simultaneous and substantial increase in agricultural production, crop yields, and employment of some 800 million peasants who depend on agriculture as their principal source of livelihood—the goal of every developing country in the world today.[8] This was achieved, moreover, without any increase in the cultivated acreage and without excluding or depriving the poor segments of the rural population of access to land or an equable share in the income and benefits of agricultural development and growth in the region. Unlike in India, there is virtually no overt unemployment in the rural sector in China today.

With an overall increase of 2 percent per annum, the rural population per arable hectare rose from about 5:1 in 1949 to 8.5:1 in 1979. Between 1957 and 1975, however, not only did agriculture employ close to 100 million *new* workers—about two-thirds of the overall increase in the workforce—but it also "substantially raised the average amount of employment for the entire farm labor force of some 310-340 million workers" by approximately 50 percent. Estimates vary. But according to Thomas Rawski, the national average input of labor had probably risen to 430 man-days for cultivating and fertilizing each hectare of sown area in 1975. In irrigated areas, "reported labor inputs range as high as 765 or more man-days."[9]

The extra labor was absorbed by various means, such as more multiple cropping wherever possible. Traditional labor-intensive practices of tillage and husbandry, like transplanting of seedlings in paddy cultivation, were now also adopted for several other crops such as wheat, maize, soybeans, rape, hemp, jute, and cotton. In cotton cultivation, for instance, each plant now receives individual attention. At the budding period fertilizer is applied directly near the root, not broadcast. The tips of cotton plants are pinched slightly and not plucked as before. In one area, farmers rely on constant plowing and loosening of the soil and timely cutting of leaves and branches to control growth during the budding period. At the final stages of plant growth, when "it is important to exterminate insect pests," the plants are pruned properly and manually weeded, "just as carefully as at the early stage" of growth. Each plant is harvested individually. Chinese sources emphasize the delicacy of this task, and according to visitors the cotton fields "are picked frequently and few open bolls are visible at any one time."[10]

The intensity of fertilizing practices has been described earlier. In the late 1960s, the accumulation, transportation, and application of manure reportedly consumed "between 30 to 40 percent of the total amount of manpower and animal power expended in the whole year."[11]

Lastly, rural works projects, like afforestation, flood control, irrigation, leveling, terracing, and reclamation of arable land, which have remained largely an empty dream in most developing countries including India, have provided gainful employment for tens of thousands of peasants in China every year. They have substantially reduced or eliminated seasonal idleness *and* achieved some dramatic results. Thus, according to a Chinese source, in the formerly grain-deficit northern provinces of Hopei, Honan, and Shantung:

Each winter-spring season, tens of millions of people braved the biting wind and snow and worked on irrigation projects. They raised and reinforced 1,000 kilometers of dykes.... Several thousand rivers and tributaries were dredged...freeing more than 6.6 million hectares of low-lying land from the threat of flooding and waterlogging. At the same time the inhabitants went in for water conservancy and other farm improvement projects, concentrating on fighting drought. Reservoirs and terraced fields were built and trees planted on the hilly areas...to prevent soil erosion. Wells and ditches were dug on the plains and alkali leached from the soil, all of which involved a tremendous amount of work. By 1970, however, the three provinces were in the main self-sufficient in grain, while their record output in 1973 was 2.5 times that of...1949, and an increase of 16,500 million kilograms over 1965.[12]

On the average, 50 to 80 million workers participate in winter-spring campaigns. In 1973-74 reportedly 6 billion cu. m. of earth was moved. As many as 110 million people were involved in that year. An estimated 32 million hectares of flood-prone lands are now protected by dikes and some 86,000 reservoirs with a storage capacity of about 400 billion cu. m. have been constructed in this manner during the past three decades. Trees have been planted to help protect the dikes. They also provide lumber. Individual peasants are rewarded for capturing animals whose burrowing weakens the dike structures. "Honan Province as a whole has mobilized nearly 100,000 peasants to capture 315,000 animals."[13]

Water conservancy is accorded an equal if not higher priority in China than fertilization of the soil. State expenditure for water conservancy is the largest single investment in agriculture—an approximate $52 billion since 1949. Construction and maintenance of tertiaries, field channels, ditches and gates, leveling of fields, and so on, are the responsibility of the communes and brigades. The labor intensity of maintenance and repair work is evident from the following instructions for limiting water seepage from irrigation channels: "loosening the dirt of the bottom and side slopes before releasing water, thus using the dirt to seal holes and crevices; tamping the soil with packers; covering the channels with clay; the use of small rocks and pebbles [that is, to seal holes]... and adding a certain amount of clay to the water to seal the crevices."[14]

Inevitably, the increase in labor intensity and workdays entailed loss of leisure. And the average level of gross output value per man-day in agriculture declined substantially between 1957 and 1975. "Indeed, with the total number of man-days lavished on a fixed land base rising since 1957 at nearly 4.6 percent annually from a high initial level, it is perhaps surprising that the average output per man-day did not decline more rapidly."[15]

On the other hand, agriculture did supply and maintain for three decades the basic necessities and food for 22 percent of the world's population with less than 8 percent of its arable land. And, clearly, it would have been impossible to create 100 million new jobs in the non-farm or industrial sectors of the economy.

In fact, if the choice in any low-income country is between work and no work for its people, as it was in China during this period and stage of economic development for one hundred million or more farm hands, the logic of increasing food production by more "efficient" or "modern" methods is far from clear. In India, for example, labor productivity reportedly remained more or less constant during the 1960s and the 1970s. Foodgrains production increased at a compound rate of 2.52 percent in 1952-65 and 2.77 percent in 1967-79. By 1979, the country had accumulated a buffer stock of 20 million tons of foodgrains and there was a general consensus that India had achieved (a precarious) food self-sufficiency, "the principal goal of our economic development."[16] In 1981-82, however, the government was forced to import 4.75 million metric tons of wheat. Once again, food had become a major national problem in India.

The rural labor force in India is much smaller than that of China—an estimated 215.93 million in 1980. But there was an overall *decline* in the number of days of wage employment for both agricultural and all rural households. In fact, the number of days rural males found work in the mid-1970s was close to that of the Chinese peasants in the 1950s. Clearly, thirty years of rural and agricultural development and technological innovation, the "prime mover of agricultural growth in a resource-scarce economy like ours," had made little or no dent on the magnitude of poverty and unemployment in the rural sector. According to the *Sixth Five Year Plan*, the proportion of casual labor in agriculture had increased along with a reduction in self-employment, indicating perhaps "the changing pattern of land holding, pressure of population on land and employee-employer relationship...."[17]

Of a total of 350 million people subsisting below the poverty level, 300 million were rural. They represented 50.82 percent of the rural population in 1977-78.[18]

Not surprisingly, therefore, even when foodgrains were available in plenty, about half of India's population could not afford to purchase that food "in spite of larger subsidisation of consumption and even a larger subsidisation of production."[19] They did not have the income or opportunity of gainful employment.

Perhaps economists and policy-makers who argue against labor-intensive techniques in agriculture on the ground that much of the increased output of foodgrains paid to low-income laboring classes employed in its production would be consumed by them,[20] should try fasting for a couple of weeks and eat only gruel for a year as most of the 800 million or more of the world's poorest people routinely do for most of their lives. It may somewhat modify their economic insights, perspectives, and priorities in agricultural development.

# SUB-SAHARAN AFRICA

CHAPTER 6

# INTRODUCTION

The goals, assumptions, and development strategies for increasing production and productivity of land and labor in agriculture in the developing countries of Sub-Saharan Africa (henceforth referred to as "Africa") have been very similar to those of India and other countries of South and Southeast Asia (henceforth referred to as "Asia"), rather than of China, Japan, and other East Asian countries.

The two regions share many common characteristics, such as colonial rule, tropical climate with the rainfall concentrated in a few months in the year and highly variable, and over 70 percent of the workforce in agriculture. Only in Ghana, Congo, and Benin is the proportion of agricultural labor less than 60 percent of the economically active population.

Again, as in Asia until 1970 or so, most of the growth in food production in Africa has been due to expansion of acreage rather than increase in productivity. And, as in all tropical areas, in the absence of irrigation there are seasonal peaks and slack labor use periods creating bottlenecks in the production cycle. The slack season coincides with the dry season after the harvest of annual crops, and is longer in the drier savanna regions.

Finally, as in Asia, there is a tremendous diversity in cultures, religions, farming systems, socioeconomic institutions, and the political economy and history of African states and tribes. Unlike Asia, however, no single foodgrain, like rice, is a common predominant crop and source of calories throughout Africa. One of a half-dozen roots, tubers, or banana plantains are the main food crops in the more humid areas; millet, sorghum, or maize predominates in the drier grasslands, in the Sudan, Sahel, and southern Africa. Rice is a major crop only in Madagascar, along the coast of West Africa, and in a few districts of Central and East Africa.

In many countries, moreover, pastoral and cash crop economies are inextricably mixed with food production. It is practically impossible in a macro survey like this to isolate the latter with regard to allocations of land and labor. *And comparative inter-country data on labor inputs per crop/hectare/year and for individual operations, such as land preparation, weeding, manuring, are virtually nonexistent.* As noted by a recent USDA report: "it is a serious matter that no data on labor utilization exist apart from case studies. Aggregating these data involves making extrapolations for ecologically similar zones and crops, inevitably producing a large margin of error, and no reliable time series whatsoever."[1]

Despite the high degree of heterogeneity in human and natural resources, however, the 45 or so developing countries south of the Sahara are generally treated as a single unit in development literature. For more specialized analysis the region is subdivided into five broad ecological zones: the Sahel, West, Central, East, and South, each containing several countries and farming systems. The high degree of heterogeneity of cultural and cropping systems and lack of relevant data preclude any meaningful inter-country or inter-regional analysis of land and labor use in the production of foodgrains in Africa. It is therefore best to examine the broad policy issues and options relating to land and labor based on resource and factor endowments; long-term objectives of development expressed by individual countries and heads of state of the Organization of African Unity in the Lagos Plan of Action in 1980; and the prognosis and expectations of donor agencies like AID and the World Bank.

## CHAPTER 7

# INITIAL CONDITIONS

## 1

Briefly, the use of land for food production in Africa falls into four categories:

   (i) Nomadic pastoralism: under which land is used only for grazing. It is not cultivated.

  (ii) Shifting cultivation: which represents the most extensive system of land use. Several crop years are followed by several years of fallow. *Less* than two-thirds of the potential cropland is cultivated annually. Shifting agriculture can even result in the gradual relocation of a whole village.

 (iii) Rotational fallow: in which more than two-thirds of the potential cropland is cultivated each year. There is no movement or shift to new virgin soils.

  (iv) Permanent cultivation: cultivation without fallow; it includes tree crops and requires fertilization of the soil.

Besides, there are two types of mixed crop-livestock farming: (a) in which animals are kept on the farm; (b) in which they are turned over to herders for grazing, a common practice in the Sahel.

Curiously, despite the rich natural and mineral resources, and established markets for exports that provide a quarter of the GDP, there is considerable pessimism and gloom about the economic future of Africa. According to the *World Bank Report*, "for most African countries, and for a majority of the African population, the record is grim, and it is no exaggeration to talk of crisis."[1] The rate of increase in GNP per person was less than in any other part of the world in the 1970s, and growth was slower in the 1970s than it was in the preceding decade.

The *World Bank Report* further points out that of the 30 countries classified as the poorest in the world by the United Nations Conference on Trade and Development (UNCTAD), 20 are African. And of the 36 countries listed in the *World Development Report, 1981*, as "low-income" (per capita income of less than $370), almost two-thirds are African.

> The tragedy of this slow growth in the African setting is that incomes are so low and access to basic services so limited. Per capita income was $329 in 1979 (excluding Nigeria) and $411 when Nigeria is included.[2]

It is true, and it is very sad. Poverty *is* tragic.

Globally, however, of the total population of 2,260 million in the low-income countries in 1979, only 187.1 million lived in the low-income countries of Sub-Saharan Africa compared to 890.5 million in South Asia. Even if Nigeria and other middle-income countries of Africa were included, the total population would still be a total of only 343.9 million.

And although Africa may have a larger number of poor *countries*, two-thirds of the *people* living in "absolute" poverty, as defined by the World Bank, are to be found in India, Bangladesh, Pakistan, and Indonesia. They number more than the *total* population of Sub-Saharan Africa. This alters the perspective somewhat, since it is people who feel and suffer the effects of poverty and deprivation—not a piece of real estate. Clearly, the magnitude of poverty is much greater in Asia than in Africa.

Again, the oft-cited problems and wasted potential due to inadequate social services, such as lack of or poor quality of water supply in the rural areas leading to ill health, lack of health facilities reducing labor productivity, the time spent in fetching water reducing time for working in the fields, absence of primary education resulting in limited access to employment opportunities in towns, are deplorable. But they are not unique to Africa.

The same is true of problems like spoilage of foodgrains in storage, fragmented holdings, low crop yields, seasonal unemployment, poor planning and implementation of development programs and projects, elitist attitudes of bureaucrats and extension agents, and, especially, preoccupation of governments "with politically more expedient short-run objectives"—a sin in which most politicians in even highly developed countries freely indulge.[3] During the 1970s, the decade of poor performance relative to the 1960s, moreover, several African economies were disrupted by wars, civil strife, political coups, and severe and prolonged droughts, that could not possibly have permitted rapid growth in any sector of the economy.

A brief comparison of the agricultural situation and achievements in Sub-Saharan Africa and India, which is often cited by experts as a model for African countries, would perhaps provide a sharper perspective.

Thus, India has twice as many people as the total population of Sub-Saharan Africa.

As in most African countries, 74 percent of the Indian workforce is still dependent on agriculture—the proportion has remained unchanged since 1960.

Moreover, since India became independent in 1947, it has had a lead time of at least a dozen years over most African nations, and of many more years over some, like Zimbabwe. No low-income country in Africa except Ghana, Liberia, and Sudan was free before 1960.

India also inherited a superior industrial, administrative, educational, research, and physical infrastructure, and it has many more trained people at all levels in the social and technical sciences, than probably any developing country in Africa.

Above all, India has had over three decades of relative political stability and an orderly planned effort to develop agriculture and industry in a friendly international environment.

Yet the per capita GNP in India in 1979 was $190 compared to $411 in Africa. India ranked twenty-second among the thirty-six "low income" countries. In Sub-Saharan Africa, only Chad, Ethiopia, Somalia, Burundi, and Upper Volta, with a total population of 55.5 million, ranked lower than India. Again, compared to 56.6 million irrigable hectares in India (1979-80), Africa has only 2.5 million hectares or 1.8 percent of the cultivated area under "formal" irrigation schemes. Of this, 65 percent is concentrated in Sudan and another 15 percent in Madagascar.

Yet the compound rate of growth of foodgrains in India between 1967-79 (the Green Revolution era) was 2.77 percent per annum, and 2.52 percent between 1952-65—the pre-Green Revolution years that would be more comparable to Africa in the 1970s. Both periods omit the two years of drought and steep drop in production between 1965-67. But although the overall rates of growth are higher in India, broken down by individual crops, there is not much difference in yields except in wheat, which is a very minor crop in Africa.

In terms of yields of irrigated rice, for instance, according to the *World Bank Report*, yields of more than 5 tons per hectare were obtained in the Mwea scheme in Kenya and the Semry scheme in Cameroon, and in several minor schemes in Niger and Senegal. In most cases, however, not more than 3 tons per hectare of paddy are achieved per hectare harvested, and not more than 2.0 to 2.5 tons per hectare cultivated. The World Bank considers these yields to be too low, even though suitable rice varieties adapted to local conditions are not yet available.[4] Actually, however, they compare very favourably with those of India and other countries in South and Southeast Asia. The highest per hectare output of rice India has achieved so far was 1.33 tons in 1978-79. The average yield of rice in South Asia in 1973-77 was 1.78 tons per hectare and 2.05 tons/ha in Southeast Asia. At least ten countries in Africa had as much or higher cereal output per hectare of cultivated land than India did in 1977-79.[5]

The most important lessons the African states could learn from the Indian experience therefore would probably be how *not* to solve problems of poverty, low per hectare output, and rural unemployment and underemployment, rather than the other way around.

## CHAPTER 8

# MAIN CHARACTERISTICS OF AGRICULTURE IN SUB-SAHARAN AFRICA

1

The alarm over the future prospects of countries in tropical Africa stems from three main and interrelated concerns:

(i) The rate of population growth. Africa has a higher rate of increase in population than any other developing region. Between 1950 and 1969, the total population in African countries increased by over 50 percent. In the decade of 1960 the farm population alone is estimated to have grown by 20 percent. It is not expected to level off for another decade or so, when, according to the United Nations, it will average about 3 percent a year.

(ii) The rate of urbanization. Less than a quarter of the population lives in urban areas, but is increasing at about 5 percent or more per year. Consequently, as in Asia, even oil-rich countries like Nigeria are facing problems of unemployment, of absorbing a rapidly growing workforce.

Urbanization is also creating a structure of tastes and demand for processed and convenient foods, such as wheat and rice, which must be imported. Their imports rose by 11 percent a year in the 1970s and accounted for 82 percent of gross cereal imports in 1977-79. The imported grains compete with locally grown staples. But the major imports are of relatively highly valued products like canned or dried meat and fish, processed milk, and a wide range of other delicacies.[1]

(iii) The rate of increase in food imports. Domestic food production is not keeping pace with the growth in population and demand.

Net commercial imports of cereals averaged about 4.3 million tons in the late 1970s. The heaviest importers are Ethiopia, Ghana,

Nigeria, Sudan, Tanzania, Zaire, Ivory Coast, Senegal, Congo, and Zambia. Food aid accounts for over 20 percent of total net cereal imports, rising from about 800,000 tons in the mid-1970s to more than 1.3 million tons in 1978. Aid was targeted mainly to countries in the Sahel and areas experiencing wars and with large concentrations of refugees.[2]

## 2

On the other hand, in many respects the region as a whole has several advantages over both Asia and Latin America, which if properly exploited could lead to a bright and prosperous future for most of the people and countries in Sub-Saharan Africa. Thus:

(i) Unlike Asia, land is not a *major* constraint yet in most African countries. Africa has more arable and permanent cropland than any other developing region.

Although the scope for future expansion of the cultivated area is declining due to the increase in population and commercialization of agriculture, land is still "plentiful relative to labor and both are plentiful relative to capital." The great majority of the peasants still has access to land.

(ii) Outside the European enclaves, low population densities and customary tenures have prevented concentration of landholdings through much of the continent. Traditional techniques have also kept the size of operational units of cultivation small. Most African farms are small and family-operated. Differences in farm size are not substantial except in Kenya and Ethiopia. Dualism is primarily the result of control of land by white farmers, as in Zimbabwe where approximately 5,500 white commercial farmers and some plantations control over 40 percent of the total land and produce most of the marketed surplus. In Zambia also, around 600 large-scale commercial farms, of which half are owned by Europeans, produce one-half to two-thirds of the marketed output of maize, the staple crop.

Within the African sector, however, even in cash crops, smallholder production is significant. It accounts for most of the expansion of tea and a significant part of growth in rubber and other export crops, such as tobacco, coffee, and pyrethrum. It has integrated an otherwise subsistence producer into a market economy. Cash received from export crops either directly or as wages earned on plantations is a major source of demand for consumer goods in the rural areas. "In many parts of tropical Africa, farm families already command resources sufficient to finance profitable investments in farming."[3]

(iii) In sharp contrast to South and Southeast Asia, there is virtually no class of landless wage workers in African agriculture. Only a negligible proportion of the labor is unemployed and seeking work. "Almost all the adult rural population participates in the labor force at some time of the year."[4]

(iv) With the exception of large farms and plantations, use of hired labor is probably the lowest—less than 20 percent—in the Third World. A major proportion of it, moreover, "is probably reciprocal labor or communal labor whereby farmers exchange labor or work in groups to perform certain tasks."[5]

(v) Although very few data are available, the consensus is that labor inputs in total agricultural production and in the cultivation of food crops in Africa are low by international standards. Rural people generally work fewer hours per year than urban workers. As in Asia, labor input varies with sex, age, climate, location, cropping system, and cultural factors. But, according to Igor Kopytoff, in the 1950s underemployment of the Suku in Zaire was so great "that even with one fourth of the men absent (from the village) those who remain suffer from essential boredom."[6]

John Cleave's analysis of farm surveys in five English-speaking countries showed that the time spent on farm operations by adult males in the early 1970s ranged from 530 to 2,135 hours per year. But all except one area reported less than 1,700 hours.[7] A nationwide rural survey in Sierra Leone in 1973-74 revealed that the average labor inputs for adults were about 1,200 hours per year. But this included agricultural processing and nonfarm work in addition to field operations. In irrigated areas, however, labor inputs were as high as 2,000 hours per year.[8]

In many areas nonfarm activities such as trading, tailoring, and blacksmithing absorb a certain proportion of the time of farm families, varying from 11 percent in Sierra Leone to 47 percent in northern Nigeria. The small industries employ labor in the slack season and agriculture provides the demand for the products of village crafts and industries.

The low level of total labor usage in on- and off-farm activities provides considerable scope, potential, and opportunity for increasing production and productivity in agriculture and ancillary industries by greater and more efficient utilization of farm labor, to "foster rapid economic growth, provide year-round employment, and reduce rural-urban migration."[9]

(vi) Finally, food is still produced by millions of self-employed smallholders who give "first priority to filling their granaries or preserving their livestock herd." The goal is to be as self-sufficient as possible.[10]

In the 1960s probably as much as 70 percent of the land and 60 percent of the labor was devoted to subsistence production. The latter still accounts for half or more of the total agricultural output. Even households engaged in cash cropping grow their own food. Consequently, demand is often thin and market supplies and prices can fluctuate widely. But, unlike the marginal and landless peasants in Asia and Latin America, most of the African *peasants* are not only socially more independent and secure from exploitation, but the threat of starvation is virtually nonexistent. At least subsistence is assured except in times of crop failures due to drought or some man-made disaster. The "skill and diligence with which they apply themselves to the task of farming are the immediate determinants of the level of agricultural output and its growth"[11]—a tremendous advantage in a poor underdeveloped country.

CHAPTER 9

# PROBLEMS AND PRESCRIPTIONS

1

Ironically, in striking contrast to the current disarray, debate, and dissensions in the industrialized countries, including the United States, over monetary and fiscal policies, there is extraordinary consensus and confidence among economists of international agencies and advisors on the monetary and fiscal measures African governments should adopt to resolve their problems of production, productivity, employment, and exports—namely, exchange and interest rates, tariffs, taxes, subsidies, and food and factor prices.

In the rural sector, trade and exchange-rate policies are believed to be "at the heart of the failure to provide adequate incentives for agricultural production and for exports in much of Africa."[1] Allegedly, they discriminate against agricultural production, favor large-scale capital-intensive enterprises, encourage imports, and stimulate migration out of the rural areas. Consequently, if the exchange rates and the incentive system are restructured and the market mechanism is permitted to operate, factor-price distortions would be corrected, resources would be utilized more efficiently, production of food crops and rate of agricultural exports will accelerate, and the farmers will receive higher incomes and become prosperous.

In fact, if the "timing" of government policies is correct, and their implementation is handled with "skill," the transition to a more commercialized system of food production would, according to the USDA report, even "avoid creation of large unmarketable surpluses and other such man-made problems.... This implies close coordination of storage and trade policies with production policies."[2] But as the obstinate agricultural surpluses, adverse trade balances, low prices for farm commodities, and high interest rates and prices of agricultural inputs and equipment in the United States clearly demonstrate, the policy choices are neither clear nor simple to design or implement. The American government has yet to discover what they should be.

In view of the fact, moreover, that the recommended reforms in trade and exchange-rate policies have to be "viewed as an instrument of long-term structural adjustment" rather than a short-term cure for balance-of-payments problems; that "the supply response to changes in incentives will take time" and can by no means be guaranteed; and that in the meantime—for an undetermined period—"these actions are likely to have an adverse effect on the overall revenue situation of governments which are already facing major problems in this regard," the caution of the African states is perhaps justified. It cannot be reassuring enough to know that "such measures *can* work, that they take time, and that change is eased and hardship reduced *if substantial* external assistance is available."[3] (Emphasis added.) At the same time, monetary and fiscal policies are not sacrosanct, and governments heavily dependent on exports, as in Africa, cannot be unaware of the constant need to review and readjust exchange rates and tariffs in response to changes in international markets and prices.

## 2

The same holds true for the increasing pressure on African states to relinquish public controls and rely on private companies and individual traders for procurement, pricing, and marketing of food crops, inputs, and services, within the country in a free competitive market.[4]

Thus, a "competitive market" requires many competitors, merchants, and entrepreneurs, and an efficient corporate and trading structure. These are scarce or nonexistent in many parts of the continent. Even if there are a few traders, one or two in each sector, the business could still become a monopoly as a result of conspiracy, collaboration, or attrition. In some countries, the only option to creating state agencies, and a cadre of native entrepreneurs in the process, may be to return to a dependence on foreign firms and immigrant businessmen who wielded a virtual monopoly of international and domestic trade and distribution of vital commodities including food in the colonial period.

It is generally overlooked that it is not possible to create "instant" entrepreneurs, like instant tea or coffee. In Europe, for example, it took around four centuries for the transformation of a class of moneylenders into pawnbrokers and then into bankers of sorts. After that it took another two hundred years (1550-1776 circa) for "commercial capitalism" and capitalists to emerge. And then, yet another century for "industrial capitalism" to get established.

Again, the managers and proprietors of Japanese corporations now striking terror in Western industrialized countries are descendants of Samurai merchants and shopkeepers who in the last quarter of the nineteenth century "understood little more than that from now on their stores were to be called 'companies'," under orders of a government trying to establish a Western-type competitive system.

Changing the *kanban* from *do* or *ya* (store) to *kaisha* (company) was simple. But they had no notion of what it meant. They did not understand *kyoso*, the early coined word for competition. Translated literally it means "running and fighting."

Not only was practically every modern industry during the Meiji era (1868-1912) initially established, managed, and financed by the state, but the government assumed the responsibility "to create new capital, nurse a new class of entrepreneurs, develop managers and engineers, train the requisite corps of industrial labor, find new markets, and, in the meantime, establish and run the new industries...."[5]

In terms of profitability, "the government factories were complete failures and a constant drain on the state finances." But it is doubtful if the subsequent pace of industrial development and commerce would have occurred "without the experience, organization, and plant contributed by more than a decade of government enterprise," intervention, and participation. The government-owned plants also faced severe problems arising from lack of experience and expertise in modern techniques, institutions, and management systems,

> ...but always with this difference: *so long as government enterprises were economically, socially, or politically useful,* it was not essential that they be profitable; and so long as the government was able to absorb losses, *there were no*

> *managerial and engineering problems that could not eventually be solved.*[6] [Emphasis added.]

It provided an invaluable opportunity for experimenting, learning by doing, and *solving problems indigenously*, a process which later became an integral feature of the private sector in Japan.


3

Another important aspect of which proponents of the free market appear to be oblivious is the apparent *necessity and cost* of regulating private enterprise even in developed industrial economies, in which the corporate, manufacturing, and trading sector is presumably *not* regarded as "exploitative and unreliable," as in many developing countries of Africa.

The United States federal government, for example, employed 81,000 bureaucrats in regulatory activities in 1979, and there were 36,487 pages of regulations "taking 127 inches of shelf space—a veritable ten-foot shelf."[7]

The estimated cost of regulating *private* economic activity was $7.1 billion in 1981, of which $1.213 billion went to agriculture. Another $1.321 billion was allocated to the Environmental Protection Agency.[8] By contrast, in estimating the investment requirements for closing the food gap in 24 countries in Sub-Saharan Africa between 1975-90, both IFPRI (International Food Policy Research Institute) and FAO provided for mechanization, fertilizer manufacture, and pesticide supply. But *no* provision was made for regulation and monitoring of the impact of chemicals on human health and pollution of soil and water that will inevitably follow.[9]

A recent World Bank study applauds the introduction of an ultra low-volume, hand-held sprayer by leading chemical companies as an "excellent example of useful innovation by private firms in Africa. These sprayers have tremendous possibilities for insect and weed control."[10] Yet, as is well known, several toxic chemicals and pesticides, which have been banned in the United States and other developed countries as a result of regulations and controls, are being used freely throughout the Third World. And they are supplied and promoted by private corporations that are fully aware of their hazards to the environment, farm workers, and consumers. Indigenous governments lack the power, capacity, and perhaps the will, to regulate their imports and usage.

In any event, the controls and "low" official prices, especially for food crops, cannot have a significantly adverse impact on production incentives in Africa because: (a) the governments lack the administrative and fiscal capacity to enforce consumer prices and ensure supplies; and (b) most farmers consume what they produce and market the surplus locally at unofficial prices. Retail controls affect mainly the sales to urban consumers. But there is also an extensive parallel market and prices are two to three times as high as the "official" price. Producers have been able to sell a part of their produce at the free market prices "and most consumers have been forced to buy at open market prices. The official prices are therefore irrelevant or only partially effective. In a large number of African countries, "food markets continue to operate, as they have in the past, without much public control."[11]

It is also pertinent to point out that in times of shortages when supplies of vital commodities, such as food and raw materials, are uncertain, as they are in most African countries, *no* developed country of any ideological persuasion permits the market to operate freely. The private sector is not trusted or relied upon to produce, distribute essential goods and services, or protect the consumers' interests.

> Specifically in markets of many sellers and many buyers—those where there is no market power—there is no mechanism by which, if there is a shortage at the going price, the supply that is available can be distributed equitably among the claimant buyers.[12]

In the last world war, even before Pearl Harbor, the Office of Price Administration and Civilian Supply was established by Executive Order in the United States. On April 28, 1942, the General Maximum Price Regulation placed a ceiling over all prices of essential and nonessential goods. Aside from regulating automobile production and civilian construction, tires, sugar, shoes, heating oil, gasoline, canned foods and meat were rationed. In Britain and Germany, "ration coupons became the decisive currency." In other words, "the invisible hand" was suspended and not allowed to allocate production resources or consumption. And the latter "was reduced not by higher prices but by rationing or by allowing shelves to go empty...."[13]

It does not mean that in a situation of shortages and a weak administration national agencies and enterprises function more efficiently than private businesses. The high administrative costs, corruption, and inefficiency of state marketing agencies and parastatals in countries like Tanzania have been extensively documented. But the choices in favor of a free market are not as clear as they are generally made out.

Finally, and very briefly, contrary to popular belief and sermons of free market enthusiasts, government regulations and controls in Western capitalist countries have not been limited to wartime or the recent decades since the last war. Even in the "good old days," in the heyday of pristine laissez-faire in America, for example, there was an astonishing range of economic regulations.

Current regulations on dairy products, for instance, date back to 1856, when with the disappearance of the family cow, Massachusetts enacted a law to prohibit the adulteration of milk, and two years later made the feeding of distillery waste illegal. "Municipal regulation of markets, of hackney rates, and of transportation of goods by wagon or dray was common in 1815." Corporation charters included not only rates and fares, but "also such matters as speed, prevention of accidents, and standards of service. Gas, water, and even ice companies came in for similar attention from the states." Regulations controlling the chartering of corporations "were of the greatest significance."

Rapid economic growth following 1815 led to a concomitant expansion of government intervention and even participation in economic enterprises either "through mixed enterprise or through full-fledged state ownership and control." Typical were the detailed specifications with regard to the size and price of loaves of bread sold in American cities. In 1818, five hundred underweight loaves of bread were seized in Charleston. In 1851, in granting a charter to the City of Davenport, the legislature of the State of Iowa provided, among other numerous regulatory powers, the right "*to regulate the weight, quality and price of bread to be sold and used in the city.*"[14] (Emphasis added.)

As far as is known, not many governments in Africa have gone that far as yet in regulating free enterprise.

CHAPTER 10

# TASTE AND TENURE

1

In the use of land and labor in agricultural production, fortunately, most African states still have a clear choice. They have three options to create a production system that will have:

(i) low productivity but a high level of employment and participation of the rural workforce;
(ii) high productivity but low participation and employment;
(iii) high productivity and a high level of employment.

No country has yet achieved high productivity with or without a high level of employment. Which of the three patterns of production and employment emerges finally will depend, however, largely on policies African states adopt in three key areas: *taste, tenure, technology*.

If the choices African governments now make in these three key areas, moreover, prove to be wrong and unacceptable to the majority of their people, within a few years, perhaps, it may be politically impossible to reverse them except by revolution and violence—as happened in China, and is happening currently in several countries of the Middle East and Central America.

As with monetary and fiscal policies, the critical question of course is which policies would be "right" and achievable in the socioeconomic and political conditions of each country.

The answer is: we do *not* know.

Contrary to the claim of development experts and observers like Uma Lele, that "the problem of Africa's rural development is not one of *not* knowing in broad terms what needs to be done...,"[1] the most important problem of aid and development has been precisely one of *not* knowing which would be the "right" policies for the modernization of peasant agriculture *anywhere*, not just in Africa.

Unlike the industrial economies that are now mainly preoccupied with efforts to "correct" the mistakes of over two centuries of haphazard growth and development, and not doing it too well, policy-makers and their foreign advisors in the developing countries of Africa and other regions have yet to identify the true nature of their problems, and often, even to ask the right questions. Their ignorance is compounded by the fact that the great majority of the developing countries in Africa, Asia, and Latin America have authoritarian, and often unstable, political systems of government. They preclude free discussion, debate, and experimentation with new ideas and approaches to development.

The key difference between modern industrial democracies and non-democratic societies is not in the level of wisdom of their policy-makers—distribution of intelligence and stupidity is fairly even among countries and cultures. Nor is it in the knowledge or access to the mysteries of the social and technical sciences. If indigenous talent should be in short supply, highly "trained" and competent scientists are available for hire and consultation. International agencies readily provide them—more readily than they provide capital.

In other words, it is not merely the lack of "trained" people "who can devise effective national strategy and policies" that is the major constraint in Africa, but the absence of institutionalized opportunities and *compulsion* constantly to review, innovate, correct, or abandon policies and projects unless and until they work, that accounts for the difference.

As even a cursory examination of economic history and record of developmental efforts over the past three decades will show, there are no models or theories yet that provide workable solutions for problems of growth with equity and institutional change in less developed, or developed, economies. A telling example of the state of the art and its practitioners is the following quotation from a report of a leading international agency on "Prospects of Agriculture in Japan" in 1955:

The demand for basic foodstuffs being very inelastic,

additional imports of over 2 million tons of brown rice equivalent would become necessary within ten years...it is at least very questionable whether Japan could increase its exports sufficiently to take care, not only of the normal increase in demand for foreign raw materials, capital goods and consumer goods, but also earn an additional U.S. $200 million which would be needed to satisfy the urgent need for food.

Adding the necessary imports of raw materials and fuel "might mean additional exports of $3-400 million."

Japan of course ceased to import rice in 1970, and it had no difficulty in increasing its exports more than to pay for its other needs for industrial and consumer goods.

The success stories in history, however, have invariably been the result of a unique constellation of several factors and events over a period of time, and are therefore impossible to replicate.

2

Of the three policy issues in Africa, namely, tastes, tenure, and technology, the first does not need much elaboration. Imports of exotic and high-valued processed foods can seriously destabilize the economy. Shifts in eating habits, from traditional staples grown within the country to cereals, like wheat and rice which must be imported in Africa, will naturally undermine commercialization of domestic production.

Because of the uncontrolled appetites of a small minority of the affluent urban dwellers, in fact, middle-income countries have already become the world's largest importers of grain and other less basic foods, such as meat, sugar, fresh fruit, and vegetables. By the late 1970s, only about one-fifth of their import bill was for foodgrains, moreover, and over one-half of the imported grain was being fed to animals in spite of widespread poverty, malnutrition, and hunger.[2] It is ironic, for example, that in spite of enormous transport difficulties and with cows all around, even in remote areas of Botswana, the shops are reportedly full of Steri-milk and Ultra-milk imported from South Africa. As observed by Michael Lipton:

The fields around Lentswe-la-Tan are beautified by sunflowers, yet its store is replete with cans of imported sunflower-oil.[3]

Donor countries share a measure of responsibility for nurturing the taste for new foods. Thus, 75 percent of United States food aid to Africa in 1978-79 consisted of wheat and wheat flour, while maize and sorghum, the staple crops in Africa, constituted only a minor share.

The solution is obvious. But it may be politically difficult. Self-interest pressures donors to build overseas markets in the commodities they want to export through the aid process. For the recipient countries, it is easy to acquire new tastes. It is much more difficult to change them. Restrictions on imports of wheat and rice have often had to be relaxed because of strong urban demand. In Liberia, severe "rice riots" of April 1979 forced the government to abandon its efforts to contain consumer demand through import restrictions and price increase. "The political repercussions of these measures cannot be underestimated."[4] China and India have demonstrated, however, that it can be done.

### 3

With regard to tenure, there appears to be a widespread consensus that "customary" rights in land are incompatible with a "modern" agriculture, economy, and society. As stated by Dr. Banda when he was Minister of Agriculture:

> ....the customary way of holding land in Malawi and the methods of tilling the land were entirely out of date and totally unsuitable for the economic development of the country...the first thing to do was to change the system of landholding and the second was to change the method of land cultivation.[5]

According to Kenneth Parsons: "New arrangements are needed to give greater scope to the expansive and liberating influences of development.... In simple fact, traditional agriculture and customary tenures in tropical Africa has no future."[6]

Reform of an agrarian structure may involve changes in one or more of three main areas: tenure or title to the land; terms of holding and scale of operation, that is, operation of the land as distinct from its ownership; and pattern of cultivation. The three sectors can be independent of each other and reform of one need not require reform of the others. In the peasant sector in Africa, however, they are probably much more interdependent than in Asia or Latin America, so that a change in the rules of title to the land will affect the operational pattern and scale of cultivation.

It is hard to do justice to a problem as complex as land tenure in such a large territory occupied by so many diverse political and farming systems.

The choices African governments make with regard to tenure will affect their policies on technology and the pattern of reorganization of various traditional systems of farming. In the absence of a cohesive land policy, however, the reverse could also be true. That is, technology could determine the de facto use of land and labor regardless of the legal forms and structure of tenure.

For reasons of historical association and dependency, the trend for most African states has been to adopt one of the Western systems of tenure together with its underlying paradigms and philosophy. The policies, however, are still in the making, and there is only fragmentary information on their impact on customary tenures. In any case, since the modern theories of economics and property originated in Europe, European experience and case studies can probably provide a far better test of their potential effect on land use than evidence and analogies drawn from other developing countries.

The emergent land policies in Africa today range from conversion of customary tenures into individualized freehold systems with private ownership to nationalization of land with a view to establishing enclaves of large-scale mechanized farms under state, private, or cooperative management in the midst of a traditional peasant agriculture based upon customary tenure.

A number of different methods of approach in creating systems of land law calculated to promote the maximum exploitation of national resources...are being tried— introduction of individual titles; registration of title (individual or communal); the substitution of 99-year leaseholds with use condition for free hold titles (Tanganyika) and so on.[7]

In Kenya the land reform took place in a capitalist framework, so that 99 percent of the agricultural output is now accounted for by the private sector consisting of small and marginal family farms and large farms controlled by individuals, cooperatives, and corporations. In the socialized agriculture of Tanzania individualized tenure has been abolished. In contrast, most African states are attempting a mix of various systems.

Thus, Sudan has three distinct sectors: participatory irrigation schemes controlled by the state; private large-scale mechanized farming under rain-fed conditions; and traditional small farms. The government of Zimbabwe is similarly seeking to establish and promote a number of production systems, namely: (a) communal farming and cooperatives; (b) private family and corporate farms of various sizes; and (c) state farms.

Most African governments still enjoy a measure of flexibility in their choice, however, because unlike Asia and Latin America, where European concepts of land, law, and private property were imposed a century or more ago, in Africa, outside the European enclaves, customary institutions regulating land use were not totally destroyed under colonial rule. In the former British colonies, "English law was never imposed...in such a way as to oust the indigenous customary laws.... It recognized the existence of two separate societies and the fact that the subsistence sector must continue to be controlled by the norms of customary law."[8] In this respect the British policy was more realistic than that of the French. As observed by David King:

> In summary, the general impact of colonial intervention on indigenous land tenure systems of Africa has been minimal. In most independent African nations traditional egalitarian distribution of rights to use land of indigenous groups has remained intact, or has been sufficiently restored, to protect the rights of the existing populations to at least a subsistence livelihood by mixing their labor with the land.[9]

More importantly, with the exception of Ethiopia perhaps, there is apparently very little popular demand yet for land reform in African countries.

4

Actually, very little is known at present about the "whys" and "wherefores" of customary tenures in Africa beyond the fact that they are an integral part of a delicately balanced farming system adapted to a particular and unique environment. Uchendu has argued that unless a government "fully understands the operation of the present tenure system and its relation to agricultural viability, it is quite risky to alter the basic principles of land tenure."[10]

Basic changes in tenure rules could also have far-reaching social consequences. According to Bohannan: "'Land reform' for the rationalization of the economy, whereby land is treated as a factor of production, means concomitant 'reform' of the social structure...."[11] In other words, land reform *per se* is not the problem. African states could compress the several centuries it took Europe to transform its feudal land systems into a decade or less, simply by legislative fiat. And there is no shortage of experts who can design and draft the most perfect reform for any country anywhere.

As stated earlier, however, given the current state of knowledge and the art of economics and other social sciences, scholars (including the author), the donor agencies, and national policy-makers do *not* really know *how* to reform a social system. They do not know *how* to create new institutions and make them work. They do not know *how* to induce controlled changes in traditional patterns of interpersonal/clan/tribe relationships, ethics of work and leisure, and, above all, in attitudes to land.

Yet these were the kinds of systemic changes, in attitudes and institutions, that were largely responsible for the "evolution" of "modern" tenure in Europe. Without such changes, even if a land reform is imposed, it will remain an empty shell or soon become the source of social unrest and disruption. And again, the experts do not know *how* to avoid the conflicts and human costs involved in the reform process. In fact, they rarely make the costs explicit, unless this helps to reinforce an ideological or political cause and bias.

In Kenya, for example, tribal exclusiveness—the unwillingness of one community, such as a tribe or clan, to allow members of another community to establish rights in lands which they regard as their exclusive domain—can and often does result in violence. It also denies "access to good agricultural land to persons who could develop and use it with their own initiative and resources." In the national interest, therefore, the Government of Kenya is advised by the World Bank further to improve, and in the course of time promote, inter-tribal land redistribution "by legal and orderly transfer *through the market*." (Emphasis added.) It is urged to take all possible steps further to develop the land market because it "offers the best possibility of overcoming tribal exclusiveness and bringing land into optimum use."[12]

The World Bank experts are obviously either ignorant or they chose to ignore the fact that *never* in history, in any region or country, has the market guided or permitted the distribution of land between different ethnic, racial, or religious groups, in a peaceful or equitable manner. Distribution has invariably been by a process of extermination or subjugation of the weaker groups. In fact, whereas the scars and memories of a war generally fade within a generation after the death of its victims, the impact and conflicts arising from changes in the control and use of land can persist through centuries as, for example, in northern Ireland.

The conflict that is now seen as a political struggle between Catholics and Protestants, originated with the establishment of an English plantation at Munster in 1583 which was followed by another and more extensive plantation in Londonderry in 1608, upon lands which had been expropriated from the Irish. The systematic policy to expropriate and drive the native Irish off their customary lands continued through the seventeenth century. By 1688, nearly 80 percent of the land was owned by English and Scottish colonists, settled on lands confiscated from Irish chieftains.

> The competition for land was such that most prospective tenants were willing to offer rents far greater than the land could reasonably produce under any system of cultivation. In consequence [the tenants] quickly fell into arrears with their payments and continued in occupation under the constant threat of ejectment.[13]

The English were mostly absentee landlords and rented their land to tenants, who again rented it to sub-tenants, the Irish Cottiers. The rents were an important source of capital in England.

The outflow of rental money from Ireland to England in 1687 amounted to £1,200,000. Since the landlords were Protestant and their Irish tenants Catholic, adherence to Catholicism became synonymous with resistance to the English ruling class on the island. And it still is three centuries later. On the English side, rather than religion, the prime motivation for the initial discriminative policies against the islanders was ethnicity, Celtic culture and Gaelic, which the English openly despised and ceaselessly attempted to root out.

The plight of the Indian tribes and other minority groups in North and South America is another example of how the weak fare under a free land market system. It is highly improbable therefore that market forces would persuade the different tribes in Kenya (and other African countries) to overcome "tribal exclusiveness" and redistribute, share, and use the land in a harmonious and equitable manner.

## 5

As a rule not much attention is paid by national governments and donor agencies to *why* and *how* a particular system of tenure evolved in a particular region or country. It is regarded as the domain of economic historians rather than that of applied social scientists.

Obviously, it is the sovereign right of each country to decide what kind of agrarian society it would like to establish. Before making a final commitment to a particular system, capitalist or socialist, however, it would be advisable for the African governments: (a) to reassess, carefully and objectively, to see if customary tenures cannot help rather than obstruct agricultural development, *the hypothesis being that there could be other, more desirable and efficient forms of land use and ownership than the so-called "modern" systems*; and (b) to examine closely, not just the current status of production, productivity, and income distribution, in the model country, but also the historical process by which it was achieved.

Thus, protagonists for individualized tenure in Africa argue in its favor on grounds of conservation and efficiency. According to them, soil fertility is declining in many areas because inalienability or lack of a permanent stake in a given piece of land constrains individual investment; communal holdings are vulnerable to the "problem of the commons."

> Where the alienation of land is strictly controlled by the group and individual farmers have only usufructuary rights in land, farmer-cultivators have at best only part-opportunities; such opportunities may very well lack the time dimension essential for long term investment in land....[14]

In a fee-simple type of private ownership, land use and investment decisions "would become more amenable to the will and needs of enterprising managers and investors, who are viewed as the agents of economic progress." The premise is that farmers "will put the land to good, or even the best, uses."[15]

Generally not mentioned, however, are the several historical examples where incentives of individual ownership did *not* result in good use, husbandry, or long-term investment in land. The most recent and dramatic example perhaps is that of the United States where incentives of private ownership of farmland have long been combined with application of "scientific" knowledge and principles to cultural practices. Yet, by 1850 a large portion of Virginia and Maryland east of the Blue Ridge had become "a waste of old fields and abandoned lands covered with underbrush and young cedars." Through the century, the American farmer generally continued to impoverish—mine—the soil until it was no longer profitable to farm. On new virgin western lands also, reports showed rich harvests for a few years. Oliver Ellsworth wrote to his brother from Bloomington, Illinois: "The soil is as black as your hat and as mellow as a[n] ash heap.... If you, John, will come on, we can live like pigs in the clover...."[16] By the 1930s, however, the fertility of that soil had vanished, probably a record in the history of agriculture. In less than a century, the Great Plains had become a *Dust Bowl*. And conservation remains a serious problem despite the several federal programs that have been in operation for over 40 years now. In major farm regions including the corn-belt states, the rate of soil loss currently averages up to 100 tons per acre per year. In many areas, such as southern Iowa, the cost of reducing the soil erosion to a tolerable level would be "three times the immediate economic benefits of doing so."[17]

Ironically, many of the soil conservation measures now being advocated in America, such as contouring, strip cropping, and minimum "low-till" or "no-till" practices with rotations that include soil-building pastures and hay are being abandoned in Africa in favor of input-intensive cultivation *in order* to avoid "environmental damage" and decline in soil fertility.

6

Historically, there is no clear evidence of a correlation between the *form* and structure of tenure and efficiency of utilization of land, husbandry, and productivity. Change in tenure will change the control and distribution of land and income, wealth and power. But it *need* not have a significant impact, for better or worse, on the traditional allocative patterns of land and labor use in agricultural production.

Nevertheless, as with assumptions regarding tenure and conservation of land, many economists and policy-makers in Africa have argued, as did the East African Royal Commission (1953-55), that individualized tenure would lead to a "commercial revolution" in agriculture.[18] They acknowledge that it would displace many smallholders. But they are confident that commercial agriculture and the stimulated urban industrial sector would absorb the landless peasants.

Curiously, like the World Bank, the Royal Commission also appears to have ignored the results of similar developments at home. Space does not permit a detailed discussion.

But before governments in Africa decide to develop a free market in land and permit individual farmers and entrepreneurs to build fences and grow hedgerows on communal land in the expectation that it will trigger a dramatic "commercial and "technological" revolution, it would be useful to examine carefully the impact of the Enclosure Acts on small farms and farmers in England in the seventeenth and eighteenth centuries.

Between 1700 and 1845, about 14 million acres—a quarter of the arable land—were enclosed in Britain. In 1883, a mere 1.43 percent of the landowners owned 87.07 percent of the total privately owned land. There was massive displacement of labor which neither the technological innovations nor shortage of labor made necessary. The majority of the peasants was driven off their lands well *before* jobs became available in the modern manufacturing (or any other) sector since this was still the era of "merchant" rather than "industrial" capitalism. Displacement broke the back of the peasantry. There were instances in which whole villages disappeared as a result of enclosure.

> All ages had been used to poverty.... But the problem of the able-bodied man without a home or a job seemed a new one.... Most trades and most towns barred their gates to the beggars.[19]

Some migrated to crowd the already crowded and disease-ridden slums of the cities. Those who stayed became tenants or landless workers. Others died miserably. Displacement created a highly inequitable and inegalitarian land-tenant system—88 percent of the land was operated by tenants in 1914. The net result of three centuries of tenurial change, in other words, was that the lower middle income farmers became poor, and the poor became paupers. Farms ceased to be the means of support of most of the rural population.

In Africa, Kenya is a good example of a country in which government intervention and efforts to create individual tenure and a market in land have been fairly effective and far-reaching. But as in the free or mixed market economies of Asia and Latin America, this has not had a significant impact on "the low productivity and continuing poverty of the mass of Kenyan people," in spite of rapid economic growth in the first decade of independence. "Kenya has not really reaped as many benefits as she should have from her impressive performance in resource mobilization and investment."[20]

The impact of individualized tenure on the *distribution* of land ownership and use, moreover, is clear and should have been anticipated. Of an estimated 1.7 million rural households, about 300,000 families—16 percent of the rural population—have no direct access to land. About 25 percent of the peasants have less than 1 hectare—less than is required for subsistence. Roughly 50 percent of the holdings are less than 2 hectares each. The bottom 50 percent of the small holdings occupy less than 4 percent of the total arable land in Kenya.

In Central and Nyanza provinces, concentration in landholdings has been increasing, and the concentration of land is greater than that for either income or consumption amongst smallholders. "For Nyanza the per capita real income of the lowest 40 percent declined by about 19 percent between 1970 and 1974, that for the middle 30 percent of smallholders rose significantly, and that for the top 30 percent increased very substantially (about 54 percent)."[21] Many of the large commercial farms, moreover, are reportedly doing poorly. They are in difficulties. Their owners who are often absentee farmers mismanage and underutilize both land and labor.

In Tanzania, on the other hand, although distribution is still skewed, and around 60 percent of the rural population is below the level of absolute poverty (in the mid-1970s), socialist policies of the government have prevented further concentration of landholdings.

Unlike Kenya, moreover, small farmers in Tanzania contribute about 75 percent of agricultural export earnings and more than 80 percent of the value of marketed cereal production. Cash crops produced by smallholders include the major tree and bush crops, such as coffee, tea, cashews, and cloves, as well as the annual crops like cotton and tobacco.

But as in Kenya, performance of the farm sector in Tanzania has been disappointing, short of expectations. The trend growth rate has barely kept pace with the rate of increase in population. With the exceptions of maize, cassava, millets and sorghum, in fact, there was a decline in the marketed output of food crops as well as of most export crops between 1970/71 and 1980/81. The average index of food production per capita (1969-71=100) for 1977-79, was 94 percent in Tanzania compared to 92 percent in Kenya.

As in Kenya, the poor performance of agriculture in Tanzania "cannot be adequately explained by limitations of the natural environment." But in neither country did the radical change in customary tenure make much difference to the low productivity of land or labor in agriculture.

7

Finally, as the recent experience of many developing countries in Asia and Latin America demonstrates, the most perfect agrarian reform, of any ideological hue, will be stillborn if the state lacks the capacity and will to implement it.

In many African countries it could be a disaster, political, economic, and social, if the state supersedes and destroys the legitimacy and powers of the traditional local groups to specify and regulate the rights of occupancy, transfer, and land use for agricultural production *prematurely*—before it has an adequate data base and the requisite organization and trained personnel to administer and enforce the provisions of the land reform nationally, and impartially.

Predictably, for a large segment of the rural population, premature reform will destroy the only economic security and guarantee of at least a subsistence living and income which peasants are assured under customary tenure, not as beggars or supplicants, but as a birthright in the use of ancestral land, without having to fill out endless forms to justify their claims.

Furthermore, the national government would have to assume the responsibility to provide food and the income to purchase the food to a larger number of people in the rural areas in addition to the small urban populace. In a food-deficit country it would inevitably increase its dependence on food imports and aid. Even in a food-surplus country, it would strain a weak internal distribution system.

For individualized tenure, moreover, preparation and maintenance of survey maps and cadastral records are essential for establishing the boundaries and registration of the land assigned to each farmer, cooperative, or corporation throughout the country. The process is time-consuming, costly, and corrupting, regardless of whether the assignment is made by bureaucrats or by local chiefs and elders. As observed by Christodoulou in 1966, the adjudication process would require:

> near faultless people, clever, shrewd, well-versed in law and custom, familiar with all details of law, custom, and practice among the group concerned, absolutely independent (and often fearless), uninfluenced and incorruptible.[22]

In heterogeneous societies the reform will open the floodgates of class and tribal conflict over land, "land-grabbing," and flight of farm workers out of the rural areas. And predictably, the rural-urban migration will occur *before* alternate jobs are or can be made available in other sectors of the economy; *before* the migrants have acquired the requisite skills to be employable in the

jobs that are available; and well *before* the state can afford to
provide unemployment insurance or income-maintenance
programs. Similar victims of progress and modernization of
agriculture over the past century, and their descendants, constitute
the hard core of the unemployed and welfare recipients in the inner
cities of America today.[23]

It is too early, for instance, to evaluate or pass judgements on the
Land Use Decree which the federal military government of Nigeria
issued in March 1978 to provide a uniform legal basis for a
comprehensive national land tenure system. But it offers a good
example of a reform that would potentially face all or most of the
above problems and difficulties in varying degrees.

Very briefly, the Decree provided for "the investment of
proprietary rights in land in the State; the granting of user-rights
to individuals; and the use of the administrative system rather than
the market in the allocation of rights in land."[24] Unlike Kenya the
land market approach was rejected, and individuals were not
permitted freely negotiable interests in land. For agricultural land,
the Decree empowered local governments to "grant customary
rights of occupancy to any person or organization for the use of
land for agricultural, residential or other purposes." Furthermore,
they were authorized to grant customary occupancy rights in such
rural lands in amounts up to 500 hectares for agricultural purposes
or 5,000 hectares for grazing purposes. And these ceilings could be
exceeded by consent of the military governor.[25] It should not be
difficult to anticipate who would apply for rights to operate such
large areas in a country where the average farm size is between 5
and 10 hectares.

Entitlement to land, even for a homesite and growing of
subsistence crops, which the peasants had enjoyed from time
immemorial as a birthright in their native villages, now became a
right of citizenship. Instead of the qualified members of the
community, the authority to issue certificates of occupancy to
specified tracts of land for a specified period was now vested in
local and state authorities, and ultimately in the federal
government. The certificates were transferable. But they could be
sold only by consent of the administration—for statutory rights of
occupancy, the state governor.

Finally, whereas every citizen was entitled to apply for land
anywhere in Nigeria, there was no guarantee that he would receive

it. That too was subject to the discretion of the public officials. The National Council of States, moreover, was authorized to grant certificates of occupancy to foreigners—"persons who are not Nigerians." Consequently, the "really significant shift in social relationships [was] not a substantive shift from kinship statuses to free contract." It was mainly a shift in the locus of power and procedure—"that is, privileges, rights and responsibilities [would] now be transferred from the individual and his kinship group to the agents of the body politic as a social entity."[26]

Since local units of government are manned by local people, and land was placed under the initial authority of local governments, it is possible that for some time at least the system of native law and custom may continue to prevail. But the whole process has now been centralized, become an integral part of the political process and administrative procedures. It can be expected therefore to suffer from the normal vagaries of politics, power, and corruption, at all levels, national, regional, and local. It would be unrealistic to expect the traditional local leaders not to abandon their social obligations and behavior sanctioned and enforced by custom, once their legitimacy has been undermined or abrogated by a secular reform or legislation.

The 1978 Decree was accepted by the civilian government which came into power in October 1979. As far as is known it still stands as the basic statement of national land policy. If fully implemented, the Decree would drastically change the status and condition of peasants in Nigeria. Not surprisingly, however, according to Kenneth Parsons, "most of the provisions seem to have been ignored. One exception to this, as evidenced by a few accounts of attempts at the modernization of agriculture, is that the governors of at least some of the states have exercised their authority to make fairly large grants of land to entrepreneurs who are willing to engage in large-scale farming."[27]

## CHAPTER 11

# TECHNOLOGY

1

The third critical problem and policy issue in African agriculture relates to technology.

As in other regions, African states perceive "modern" technology as the panacea and key to agricultural development and the creation of a modern society and nation; an instrument for increasing man's control over physical nature.

Although a variety of new, mainly export crops has been successfully absorbed into the traditional farming systems, "African agriculture has probably been less affected by technological change in the past 20 years than agriculture on any other continent."[1] Land use methods have been modified in many areas and shifting cultivation replaced by recurrent tillage—a rotational system of fallowing which allows the land to recuperate its natural fertility. But with regard to the nature and pace of future changes in the organization and scale of agricultural production, new yield-increasing inputs, and tillage and harvest equipment, African governments still enjoy a wide range of choices.

Nevertheless, as with land tenure, most states appear to have opted for the Western model of capital-intensive agriculture. The choice stems from a conviction that peasants using traditional techniques on small holdings cannot be the basis of a modern agriculture; the belief that "only a rapid transition to mechanized, high productivity schemes, as practiced in the industrialized world, would overcome the stagnation linked with the traditional low-input, low-output methods." Such technology is also considered a reasonable solution to labor shortages, where these exist.[2]

Many Western economists and aid agencies believe that

transition to a permanent, land and input-intensive agriculture is an imperative, both for increasing commercial production of food for the growing populations and for preventing further deterioration of soil structure and fertility—that, in effect, there is no real choice between traditional and modern technology in Africa.

Failure to make this transition successfully will inevitably mean both a deterioration of the natural resource base and an increase in rural poverty.[3]

The arguments underlying the belief are: (a) that technical improvements based on traditional knowledge and inputs yield only marginal returns. They would therefore not be readily accepted by farmers. The gains must be substantial for a technical package or innovation to be widely adopted; (b) that the needed increase in production cannot be obtained except through "new high yield technology and large increments of purchased inputs."[4] Presumably therefore neither the Chinese policies in the two decades following the revolution, nor the equivalent of the agricultural revolution in England in the eighteenth and early nineteenth centuries in the pattern of land use—the so-called "new husbandry"—would be appropriate because they "involved little more than the final destruction of medieval institutions and the more general adoption of techniques and crops which had been known for a long time."[5] The inputs were largely supplied by the agricultural sector. The only qualification and concern that both donors and aid recipients share is that the new techniques should be "appropriate," adapted to local factor endowments and socioeconomic conditions.

As in the case of tenure, not much is known about the traditional farming systems in Africa. But because "appropriate" techniques for the various farm systems have yet to be developed, or even defined; or because those that are available, such as contour ridging, sowing at the right time and depth, spacing of plants and mixed cropping, and use of farmyard manures to increase soil fertility, are not regarded as "modern," the policy in many countries has been to go ahead and import the latest equipment and inputs. The exporting countries provide the intellectual and economic justification and feasibility studies for doing so, together with technical advice, consultants, and, often, the credit to purchase the hardware. "In fact in a number of countries the availability of suppliers' credits has at times governed the types of equipment and machinery procured without much reference to the specifications that really should have been met."[6]

Furthermore, because large machines cannot operate efficiently on small scattered plots of land, and peasants do not have access to adequate credit to purchase manufactured inputs, it is considered necessary to make "institutional arrangements which facilitate the achievement of economies of scale in production, the handling of input supplies and the marketing of outputs, and the provision of economic, social and recreational services."[7]

To enable the subsistence sector to adopt modern techniques and benefit from economies of scale, the "arrangements" include various types of settlement schemes involving relocation of peasant holdings and peasants into large consolidated blocks of arable land under some form or degree of cooperative management. Tractor-hire service and inputs, seeds, fertilizer, and other chemicals are also being offered to small farmers at heavily subsidized rates. In the commercial sector, which is expected to provide the bulk of the future growth in marketed food and export crops, on the other hand, the trend has been to establish large lavishly mechanized farms under private or public ownership, often in partnership with multinational corporations or under the management of hired expatriates.

As can be expected, the commercial farm projects have grown at a faster pace than those in the subsistence sector. But in either case, in areas where, until a decade or so ago, tools and techniques predated the plow by animal traction in Egypt some 5,000 years ago, and village artisans could not make or repair a cart wheel, African agriculture was and is being expected to make a direct transition from the hand hoe and machete to the tractor and combine harvester.

In the 1960s, the total capital investment in a plow, cultivator, cart, and a pair of oxen required less than $300 compared to $4,000 to $5,000 for a tractor with a minimum number of implements. But the "limitations of animal-drawn implements, the impatience with the problems posed by their use and the widespread desire to embrace immediately the technology of the twentieth century produced numerous schemes for mechanizing agriculture" in countries such as Kenya, Sierra Leone, Ghana, Zambia, Gabon, Lesotho, and Nigeria among others.[8]

In most countries, moreover, even the animal-drawn equipment—plows, cultivators, and carts—had to be imported. Ox carts were the most expensive item, the equivalent of $100 or more in the mid-1960s, because African artisans had not "been trained to make the wheels which are the most costly part of the cart."[9] In Ghana, when government policy switched to mechanization of clearing, plowing, and reaping operations in rice and maize production in 1962, small farmers like those in Bawku, where the number of ox-plows had increased from 497 in 1952 to 2,645 in 1960, found it extremely difficult to have "their ox plows repaired and impossible to replace them."[10] By 1980, therefore, even in areas where it is possible to raise livestock, animal traction—the logical intermediate step between manual and mechanical draft power—had been adopted "only here and there."[11]

According to the 1970-71 census of agriculture in the Northern and Luapula Provinces in Zambia, there were 201,000 farm households and only 4,700 trained oxen. "The handling and training of oxen is almost as foreign to villagers in the Copperbelt, Luapula, Northern and Northwestern Provinces as is the operation and maintenance of tractors."[12] The government sought the remedy for low productivity and rural poverty in mechanization.

> Tractors were to be supplied universally. When they all broke down in 1965, the Government sought its remedy in credit. And as the millions of Kwacha slipped away, the nation turned to cooperatives in order to enable villagers to combine their efforts so as to achieve the scale of operations that would make tractors and credits worthwhile. The cooperatives, of course, were formed in great numbers, particularly where agriculture was most backward.... But by 1970 it was obvious...that cooperatives had got off to a very bad start, and most were closed down either of their own accord or by the Department of Cooperatives.[13]

In other words, many African states have chosen to introduce new agricultural techniques and equipment that are capital-intensive, costly, and must be imported because they are not manufactured domestically. Because they are so expensive, they require heavy subsidies in order to make their use economic and acceptable even to the more prosperous farmers who are generally the prime, and often the sole, beneficiaries.

Policies and rates of subsidies vary between countries and only limited information is available. But in Niger, for example, the average subsidy rate in 1978/79 was 54 percent of the cost for fertilizer, 59 percent for pesticides, and 77 percent for animal traction.[14] Nigeria subsidized tractor-hire service up to 70 percent of cost. In Ghana, mechanized services for land clearing, cultivation, and harvesting of rice received a subsidy of up to 78 percent; and fertilizers between 24 and 63 percent.[15]

The government in Sierra Leone started to provide tractor plowing services to farmers in 1949. It maintains a fleet of tractors which "plows, harrows, and sometimes seed harrows for farmers who pay a highly subsidized fee, presently Le24.70 per hectare. This service costs the government about Le108 per hectare."[a] Subsidies on fertilizers (in 1975/76) ranged from 57 to 66 percent. Because of the heavy subsidies it has been difficult for the government to provide adequate funds for importing the needed fuel, spare parts, and new tractors, and sustain the same level of services to farmers. The area mechanically plowed has fluctuated widely, "reaching 11,250 hectares in 1971, dropping to 8,000 in 1973 and increasing to 21,000 in 1974, then dropping again to less than 10,000 hectares in 1977."[16]

According to an FAO estimate, mechanization would require an investment of $5,153,000,000 (1975 prices) between 1975-90 for closing the food gap in 24 select countries (less Sudan) in Sub-Saharan Africa. Cost of fertilizer inputs alone is estimated at $2,328,000,000 in U.S. currency annually by 1990.[17]

In addition, large investments are needed to improve and build new food storage and processing facilities, roads, transportation, and information infrastructure to enable the delivery of new inputs and techniques to farmers in different regions and of their surplus produce to urban markets and consumers. Senegal had to seek financial assistance for transporting domestically produced grain from surplus to deficit areas within the country.

Not surprisingly, therefore, and contrary to the proclaimed goal of self-reliance, the problem of funding imports of food and

a. Le1.00 = $1.00.

technology has been acute in most of Sub-Saharan Africa. In 1967, for example, the Arusha Declaration had affirmed that "Independence means self-reliance." As explained by President Julius Nyerere, it is stupid "to imagine that we shall rid ourselves of our poverty through foreign financial assistance rather than our own financial resources.... Tanzanians can live well without depending on help from outside if they use their land properly."[18]

A decade later (1977), 60 percent of the development budget of Tanzania came from foreign aid. Gross aid disbursements between 1973 and 1980 averaged around US $350 million per year. Six donor countries wrote off outstanding debts in 1978-79 and converted all new assistance to grants. Yet the outstanding debt from aid and nonaid sources amounted to $1.25 billion at the end of 1979. Furthermore, 90 percent of the landrovers, over 30 percent of the trucks, and 40 percent of the tractors were reportedly off the road in 1979 for lack of maintenance. Shortages of trained staff plagued the administration at all levels. There were only 100 trained accountants, for instance, and over 300 parastatals, the agencies of government operation in the economy, were in need of bookkeepers.[19]

In the region as a whole, current account deficits rose from $1.5 billion in 1970 to $8 billion in 1980. External indebtedness climbed from $6 billion in 1970 to $32 billion in 1979. "Fiscal pressures also intensified in many countries, as indicated by declining real budgetary allocations for supplies and maintenance, growing imbalances between salary and nonsalary spending, and difficulties in financing local and recurrent costs of externally funded development projects."[20] Even an oil-exporting country like Nigeria was forced to cut off nearly all imports in an effort to stave off financial collapse. Its external debt had grown to an estimated $9.3 billion at year-end 1982.

2

Curiously, the bias and preference for large-scale mechanized agriculture persists in Sub-Saharan Africa despite the fact that the new methods have not had a significant impact on land and labor productivity. In no country, "even Kenya, has there been a clearly visible departure from the production trend line that could be attributed to the adoption of a package of inputs based on new technology."[21]

It is significant, moreover, that the stagnation or decline in crop output and labor productivity in the 1970s occurred despite the fact that the various governments and donor agencies were focusing on agricultural development. Sub-Saharan Africa received a total of $5 billion in aid for agriculture between 1973 and 1980—most of it for the food sector. By 1981 this aid was running at the rate of approximately $9 billion a year. In many countries grants and low-interest loans constituted over half of the investment in rural and agricultural development.

The donor-financed projects, in fact, have had very limited impact. "This holds," according to Uma Lele, "irrespective of whether their achievements are judged by inputs such as numbers of local and expatriate staff recruited, research trials carried out, amounts of fertilizer and other inputs distributed, vehicles purchased, buildings and roads constructed or maintained, or amount of data collected or analyzed by evaluation units, or by the end results such as increases in yields..." and output.[22]

Total net imports of cereals increased from 3,776,100 metric tons in 1975 to 5,484,100 tons in 1979. Instead of eliminating their dependency on food imports, moreover, both the food deficit and self-sufficient and surplus states in Africa have acquired another, even greater dependency—on increasing, and often unreliable imports of costly chemical fertilizers, pesticides, and herbicides; often inappropriate and expensive farm equipment and motor vehicles; and, in many cases, the fuel without which the machines will not work. In Kenya, for example, in the early 1970s some of the extension staff could operate "only the first two months out of a six-month budget period due to lack of fuel for their vehicles, and the situation in Kenya was better than in most other countries."[23]

Aside from fuel of course it takes a high level of mechanical and management skills and experience at various levels to run large mechanized farms efficiently. Many large-scale farming projects, such as state farms and land settlements, and tractor mechanization[a] and tractor hire service schemes have generally

---

a. Tractor mechanization is defined as relatively large tractors—40, 50, 60 hp— and associated equipment.

failed after a few years or are operating at far less than the optimal degree of efficiency and utilization in Africa because of poor management and a shortage of trained mechanics. Costly sophisticated equipment often becomes corroded and is reduced to scrap prematurely due to poor maintenance and lack of spare parts which too are costly and must be imported.

Agbede and Warrake, two of the largest and oldest fully mechanized state farms in the Bendel State of Nigeria are a good example. They were established in 1972 and 1974, respectively, "...out of a realization that peasant farming alone could no longer solve the problem of food production in the face of increasing population growth and rising food costs."[24] The projects attracted many distinguished visitors from other countries in Africa and abroad.

Very briefly, the state acquired 6,000 hectares at Agbede and 5,200 hectares at Warrake. The goal was fully to develop 4,000 hectares for cultivation on each farm. Both farms were heavily mechanized from the start. By 1976, however, many of the machines lay idle for lack of spare parts or repair. In Warrake, in fact, there were no maintenance facilities on the farm. Lack of adequate supplies of seeds, fertilizers, and other inputs, at the right time, was also a major bottleneck on the two farms.

Over a period of five years, neither farm made any profit. The harvested acreage shrank instead of increasing. The target of 4,000 cropped hectares was never attained. And there was a steep decline in yields of the two principal crops—rice and maize. Both projects did very poorly.

Agbede and Warrake are located near each other in the same ecological zone. In the rest of West Africa, however, it is difficult to compare productivity and yields, especially of rice, because it is grown under very different natural and technological conditions. These range from traditional rain-fed upland, to swamp cultivation with no modern inputs with long periods of fallow, and to intensive mechanized cultivation under total water control.

Labor inputs per crop/hectare also vary enormously between and within countries. The variations appear not to be very closely linked with the degree of water control, and only slightly with differences in yields which tend to be fairly similar for each technique. But labor time is positively correlated with mechanization which unlike irrigation has very little impact on per hectare output of rice. There is also a fairly strong correlation between yields and extension costs probably because these costs are associated with the delivery of modern inputs.

Thus, according to a survey conducted during 1974-75 in Sierra Leone, tractorization of land preparation in the Bolilands reduced labor inputs from 317 to 193 man-hours per acre. The average size of farms increased. But returns per unit of land did not improve. Nor was mechanical technology "particularly successful as measured by returns to labor."[25] In fact, if the heavy subsidies on tractor-hire services are discounted, hand cultivation is considerably more profitable than mechanical cultivation. Impact on national production has been quite small. In Ivory Coast, Liberia, and Senegal also generally advanced techniques have failed to improve the efficiency of rice production.

In Mali, on the other hand, traditional techniques predominate in rice cultivation. Except for mechanical threshing of part of the crop, there is no mechanization of field operations. Rice is harvested manually, with a sickle. Fertilizer is used sparsely, only under irrigated conditions. Labor costs per kilogram of rice are lower in Mali than in Sierra Leone, Ivory Coast, Liberia, and Senegal. Wages in Mali are fairly low compared to other countries, especially Ivory Coast, but labor productivity is relatively high. In addition, "Mali clearly has the highest rates of net social profitability of any of the five countries."[26]

The earlier policy of mechanizing rice production and establishing collectives and state farms was abandoned in Mali in 1969. Customary tenure was restored. But the government continued to control the disposal of about 50 percent of the marketed produce and the purchase price has been relatively low. Private trade in foodgrains is banned. All agricultural inputs are sold through the Societe de Crédit d'Equipment Rural (SCAER) which can be assumed to function with the normal inefficiency of a state agency. Subsidies on farm equipment sold by the government also were removed before the 1977/78 crop year, and the government announced its intention to remove all remaining subsidies, including those on fertilizers.

The policy has been mainly to promote, subsidize, and invest in improved seeds, extension services, and, above all, in the improvement of the traditional uncontrolled flooded system of rice culture. The latter is being replaced in Operations Riz Segou and Mopti by improved but relatively inexpensive techniques which provide limited water control in diked polders. The system consists of an unleveled polder with an inlet gate, a common canal and drain, and an earth dike encircling the cultivable area. Empoldering enables the regulation of the rate and timing of flooding of paddies and retention of the water in the fields.

Introduction of low-cost water control into the traditional production system has increased the average per hectare yields of rice in Mali from 0.836 metric tons in 1969 to 1.116 metric tons in 1976. Production went up from 135,000 tons to 250,000 tons in the same period.[27]

The government's ability to expand irrigation or empoldered area depends, however, almost entirely on external aid. But compared to Ivory Coast, Senegal, Sierra Leone, and Liberia, Mali is a much poorer country, with a poor rural infrastructure and a low level of higher education. And unlike Ivory Coast in particular, which hires a large number of foreign technical experts in spite of having the highest level of advanced education, Mali does not depend significantly on foreign talent to improve the production and technology of growing rice. Nevertheless, Mali has succeeded in becoming self-sufficient in rice. In good years it even exports the grain.

In most countries, however, little or no attention appears to have been paid to the questions of (a) whether "modern" inputs and equipment are absolutely essential or desirable for increasing land and labor productivity and employment in agriculture; and (b) if they can afford them or use them efficiently given the average level of education and skills of their farmers.

As was found in the case of the Green Revolution in Asia, it is not sufficient merely to substitute one seed for another and throw a ton of chemical fertilizer on the soil. Hybrid SR52, heavily fertilized on soil deep-ploughed by tractor, produced ten times the yields of traditional maize varieties in Zambia. But as observed by Christie and Scott, the impact of the new technology on the traditional farmer was equally devastating. He simply could not cope "with the heavy demands it placed upon financial, mechanical and computational skills, as well as upon sound judgement and energy.... Unless he is highly numerate and literate he is heavily dependent upon outside help for such minor details as the calibration of his fertilizing and seeding equipment (and the repeated checking of the same). He is caught in the early planting quandary; he needs constant advice on pest and nutritional problems, and, by no means least, he must adopt a far from traditional attitude towards money." And of course the extension service that might have provided the needed help with his management and technical problems is "far from equal to the task."[28]

On the other hand, according to agronomists Kirkwood, Brams, and Chang, in Sierra Leone a single early weed eradication could increase rice yields by perhaps 5 percent from the present level of 1,200 kg/ha. It would entail only minor modifications of the farmers' work patterns. In West Africa, subsistence output of rice could increase at a rate sufficient to improve the people's diet and override the annual population growth of 2.3 percent. Only simple innovations based on traditional cultural mores and indigenous resources would be required. Similarly, yields of 2,400 kg/ha could be achieved under the present shifting agriculture systems, if the fallow period is not curtailed too severely—under five years.[29]

An average yield of 2,400 kg of rice per hectare would be no mean achievement. It would be higher than that of most countries in South and Southeast Asia today, over a decade after the Green Revolution. But it would be approximately equal to the yields achieved by peasants in other agrarian societies, as of Egypt, China, and Japan, *prior* to the introduction or any knowledge of the scientific method and technology.

3

As demonstrated by the performance of high-yielding varieties of rice in South and Southeast Asia, problems of management and technological inefficiency in the utilization of modern agricultural equipment and inputs are not unique to the countries of Sub-Saharan Africa.

Unlike the seed-fertilizer technology that initiated the Green Revolution in Asia, moreover, except for a few *innovations* like the development of hybrid maize, viable *packages* of high-yielding seeds, inputs, and agronomic practices have yet to be developed for a comparable "Black Revolution" in Africa.[30] And the basic resource data that would be essential for the development of appropriate technical "packages" for the various ecological and cropping sub-systems, such as rainfall patterns, physical properties, and water retention capacity of soils; quality and quantity of vegetation cover; river flows and drainage patterns; are limited or nonexistent.

> Africans and donors alike know far too little about the science of tropical food production. Nor have we had any more success in converting the little we do know into production increases.[31]

The hazards of designing and implementing new agricultural techniques and untested innovations are obvious, especially in locations with fragile soils or erratic rainfall. In the field of crop technology, the crop-specific approach which produced the high-yielding varieties of rice and wheat that could be adapted widely in Asia has failed to produce similar results in Africa "in part because of the intractability of the crop adaptation problem, in part because such packages need to be tailored to the labor scarcity conditions of African agriculture."[32]

Capacity to provide backup research to support and maintain higher levels of productivity of improved practices and plant varieties that are available also is very limited, "particularly when new technical packages are introduced from beyond Africa."[33] In fact, even the methodology for tailoring technical packages to fit location-specific requirements has yet to be developed. "The scattershot approach to finding solutions to real problems has produced no significant results in farmers' fields for a painfully long period of time, while food gaps in African countries have grown larger."[34] Development of "effective technical packages" for each region and farming system therefore could take not a few years but decades.

Under the circumstances, it appears to be somewhat premature to expect or highlight the importance of not only "localized testing," but *"fine tuning* of technical recommendations...to take account of local physical and socioeconomic conditions to establish relevance and acceptability." Thus, aside from genetic modifications to meet localized taste preferences, according to the World Bank, general "fertilizer recommendations may need fortification to meet localized micronutrient deficiencies, and so on."[35] (Emphasis added).

Given the severe limitations of human and fiscal resources, physical and administrative infrastructure, and lack of information and data about various farming systems, obviously *any* technology that requires such fine tuning to local conditions would be both inappropriate and unworkable. The strings will snap in the process.

# CHAPTER 12

# RESEARCH AND PARTICIPATION

1

What then are the options?

> The range of choice is bounded by technical knowledge about the productivity of various combinations of inputs, while the rate and directions of change are determined by the ability and willingness of African farmers and farming communities to alter existing patterns of production.[1]

Therefore, instead of waiting for scientists to develop appropriate packages of "modern" technology and inputs for each farming and ecological system, which the peasants may or may not adopt, perhaps the most pragmatic strategy for the present would be to entrust the prime responsibility for innovation and research to the local farm communities to improve the efficiency of tested techniques. At present they alone know and understand their particular environment; why they are doing what; and how much more they could do with their labor and local resources to increase their output—per crop/hectare/year. Instead of pouring scarce capital into largely unproductive and understaffed national institutes and research systems, described as "engineering equivalents" of teaching hospitals, therefore, why not invest more in the *peasants* who produce the crops?

The investment could be accomplished by decentralizing research by establishing low-cost village research centers in key agricultural areas. These would be linked to a provincial center which in turn would be linked to an apex research institute at the national level. The program would be administered separately from other service and rural development programs, such as community development, adult literacy, or basic needs.

This is a broad outline for a strategy and not a blueprint.

The most important component of the structure would be the *village* center. It would provide the institutional framework for generating the initiative and new ideas for using land, labor, and local resources more efficiently or differently at the grassroots level. This would be distinct from and in addition to the conventional role of research and extension of merely transmitting distilled knowledge about modern techniques to the peasants.

Specifically, the role of the village research centers would be fourfold:

(i) to encourage, involve, and work with the bright and enterprising peasants—women in areas where they are the main producers—to think, discuss, and experiment constantly with new ideas and cultural practices, such as crop rotations, soil and water management, seed selection, manuring, weeding, and so on;

(ii) to initiate experiments with new designs of farm tools and equipment and generally upgrade the level of skills in the community for the making and repair of plows, carts, harnesses, water pumps, and so on;

(iii) to test promising innovations and convey information about them to the provincial center and obtain guidance for technical problems that the village center is unable to resolve;

(iv) to test new inputs, seeds, or practices that the provincial center might recommend with the full involvement of the local peasants before adopting them. The testing should be done by the farmers and not itinerant agronomists from national or international centers.

Aside from any research they may initiate, the main role of the national and provincial centers would be to provide support and coordination for testing and diffusion of new techniques, local and imported, throughout the country. Thus, reportedly over a dozen effective indigenous methods of crop protection are practiced by the peasants in Southeast Nigeria. It would be far more useful to test and promote their use widely than invest indiscriminately in pesticides.[2] The national and provincial centers would also have the responsibility for funding, recruitment, and training of the staff of the village research centers. Preferably, it should be small and recruited from the local community.

The key person at the village level should be an agriculturist with a maximum of high school diploma, and a lot of *common sense*, but familiar with the basic scientific principles of agronomy, soil science, plant pathology, entomology, and animal husbandry, as well as with the indigenous crops and practices. He should be assisted by a local carpenter and blacksmith with training in elementary mechanics.

It would be necessary to ensure that the training the research and extension personnel at the village, province, and national levels receive is appropriate, relevant, and useful in the rural environment of the country. And it should not be limited to the educated elites. For every student sent abroad for a degree in the agricultural sciences, at least a hundred or more peasants and village artisans should receive training in the basics of their trade.

In areas where traditional community institutions are still functioning it would be necessary and desirable for the village research center to work closely with the local council and leadership.

The unique feature of the system, as distinct from the current extension education system, would be its involvement of peasants as *equal* partners in the *research process*. They will not be merely passive, though "rational," targets or recipients of "superior" knowledge from people "who have had the privilege of higher education."[3]

It would also reduce the various steps of the process—the lag between the birth of an idea, testing, and, if found useful, translation of the idea into a new tool or practice. The probability of the innovations being viable and appropriate would be very high because (a) the peasants do not have the expertise to make them otherwise; (b) unlike experts in distant laboratories, they are less likely to overlook the obvious or scorn simple solutions.

There is no need moreover for the developing countries in Africa (and elsewhere) to invest time, money, or energy in basic research at the present time. This should be left largely to international research centers and the industrial nations. In the applied field, moreover, the policy should be to import mainly the knowledge and principles embodied in a technology; not the finished product of that technology.

Thus it may be necessary to import the information on why tropical soils need nitrogen or what might be the most efficient method of applying it to crops. But it is not necessary to import the fertilizer since chemical nitrogen is exactly the same as the nitrogen from organic sources, like oil seeds, and animal, human, and natural waste materials, which can be produced and processed locally. Application of *knowledge* rather than of the *product*, is perhaps the most crucial ingredient of the learning process in technological development. And it cannot be learnt by proxy.

The search for knowledge moreover should not be limited to current techniques in industrial countries. The practices that prevailed when they were at the same stage of development as the developing country is in now would probably be far more relevant and useful.

Thus, lack of rural roads and transportation facilities is a major problem in most African countries. In Zimbabwe, "Commissioners wishing to visit peasant settlements had to use four-wheel-drive vehicles; in other areas roads were impassable."[4] In designing its future roads, however, it would be far more instructive for the Ministry of Roads in Zimbabwe and other countries to study the road and transportation system, for internal commerce and travel, in the United States in the nineteenth century rather than in 1982. The latter incidentally was not only very costly to construct, but is now in a serious state of disrepair. Nor is the automobile the most efficient or the only feasible mode of transportation. Large transcontinental empires were conquered and administered long before the invention of the steam engine or the motor car in the eighteenth and twentieth centuries respectively.

There is no guarantee, of course, that decentralization of research will indeed produce pathbreaking innovations in a given number of years. But then, neither does the present system. Much would depend on the environment: (a) whether the peasants perceive the need and urgency for more efficient use of land and labor in crop production; (b) whether they believe that it is indeed possible to increase production with local resources; (c) the respect and spirit with which their suggestions are received and recognized.

Decentralization would also involve a radical change in the concepts and attitudes of the educated elites, scientists, and administrators towards peasants, agriculture, and native institutions and traditions—a narrowing of the social and intellectual segregation that divides them. But to deny the potential and possibility that peasants can and do innovate would be to deny 5,000 years of history of the development of agricultural technology prior to the nineteenth century.

Eventually, in Africa more sophisticated research and off-farm inputs and equipment will become necessary to expand further the potential of land and labor productivity in agriculture. But, if the investment in education and infrastructure is adequate, the peasants would have acquired by then the requisite skills to use more complex tools and techniques, and the state would have developed the needed resources, means, and capacity to deliver them to the target groups.

To reverse the sequence would be to put the cart before the horse. Making modern capital-intensive inputs and equipment available prematurely to select individuals and entrepreneurs, a privileged minority of large farmers, on the assumption that they "can be used to spearhead the introduction of new methods."[5]

The only possible way to protect peasants from exploitation and forced eviction from the land would be to consign them to fenced-in reservations and settlements and to a status of perpetual inferiority, exactly as the colonial powers did. At the same time, there can be no assurance that the "modern" or large farmers will indeed use the land efficiently and maximize production. It is highly improbable, in fact, that they will. Internationally, the current strategy will perpetuate the very dependence—intellectual, economic, and political—that the African states are trying to escape. But then of course, as observed by Frances Stewart:

> There is a complex system of relationships between past policy towards technology, policy makers and policy making. Countries which have been heavily dependent on foreign technology, particularly in the form of foreign investment, find it more difficult to regulate it.... Similarly, patterns of production and consumption which are broadly inappropriate set in force strong forces making for similar inappropriate choices in the future....[6]

There is an obvious contradiction in the affirmation by the

Heads of State and Government of the Organization of African Unity that "Africa must cultivate the virtue of self-reliance," and at the same time call upon international agencies to provide 50 percent or more of the investment requirements in agriculture between 1980 and 1985. They deplore the lack of will of the developed countries to provide more resources for "accelerating their development."[7] Unfortunately, it is they, the African states, who have lacked the will and confidence to rely on their own resources and ingenuity.

# LATIN AMERICA

## CHAPTER 13

# LATIN AMERICA

Problems of land and labor usage in agriculture in Latin America do not fall strictly within the scope of this book.

Except for Haiti, with a population of 4.9 million, no country in Latin America had a per capita GNP of US $370 or less in 1979, as did the 35 low-income countries in Asia and Africa. And even Haiti was more prosperous than India, Sri Lanka, Sierra Leone, Malawi, and several others in those regions.

Latin American countries are in the "middle income" category and, regional diversities notwithstanding, according to the World Bank, they share two common characteristics as distinct from the low-income countries: "their growth prospects are more sensitive to economic conditions in the industrialized countries...and they have more resources to raise the living standards of the poor."

Latin America has more cropland per person in agriculture than Africa and Asia. The countries tend to be more urbanized. And of the total population living in "absolute" poverty in the rural areas of the developing world in the mid-1970s, Latin America and the Caribbean countries accounted for only about 4 percent.

Plantations moreover are excluded from consideration in this book. And unlike Asia and Africa, the plantation is the dominant mode of production in Latin America. It is also the nucleus for employment of the peasants in the surrounding area. "Organisation, economic activities, etc. revolve around servicing this large estate."

Even in countries in which the traditional landlord patron has been replaced, the new estate operator controls capital and access to public goods and services for the rural poor. As noted by Andrew Pearse,

The patron in land, eliminated by land reform, gives way to

the patron in capital or authority, and the dyadic relationship between this new patron and his peasant clients gathers up the cultural traits of the old relationship and becomes a basic principle in the new ordering of social relations.

The single most important issue in the hemisphere therefore is land control and the power that goes with it. Unless this is altered substantively it would be impossible to change the organization and efficiency of land and labor usage in agriculture. That could be achieved, however, only by the political process.

Even the obvious remedy of increasing the intensity of land use in areas where small farmers are concentrated would not be feasible according to major international agricultural research institutes such as CIMMYT (International Center for Maize and Wheat Improvement) and CIAT (International Center for Tropical Agriculture). According to CIAT:

> Data from rural development projects in the highlands of Mexico, Peru and Colombia suggest that agriculture is often only a minor part in the total income of subsistence farm families. Furthermore, the situation of small farmers in the highlands is often one characterized by limited land and a lack of well-developed infrastructure....
> The role of new agricultural technology in promoting the welfare of these rural poor is limited...the agriculture of the highlands has evolved over long periods of time on relatively infertile soils and is believed to be operating at a level near its potential...the probability of substantially increasing food supplies through new agricultural technology is low.

On the other hand, the inevitable inability of the campesinos to take advantage of commerical opportunities due to lack of management skills, knowledge about and access to new techniques, and markets and credit, in a dual structure of production is also a major factor and cause of their backwardness and poverty.

Thus, some Mexican *ejiditarios*—small farmers who received land through agrarian reform—have land in irrigated districts where cash crops like cotton, fruit, and vegetables are cultivated. But a 1966 survey "of *ejiditarios* with individual tenure in a large irrigation district in the state of Michoacan showed that less than 32 percent of the *ejiditarios* worked their own land: 51 percent rented their land to private entrepreneurs who used it to grow cotton; and 17 percent who left the land fallow. The projected trends were that by the 1970s, 81 percent of the *ejiditarios* would rent their lands out and only a minuscule number would actually continue to operate their parcels themselves."

Finally, there is practically nothing in the experience of colonial history or recent developments in the low-income countries in other parts of the world that is relevant to Latin America. But in countries in which land policies are still in the stage of being formulated, as in Africa, or where they could still be modified, even reversed, perhaps, without tearing apart the social fabric, as in Asia, it would be instructive for the policy-makers to study the situation in different countries in South and Central America to comprehend fully the consequences and impact of a dual structure of land ownership, use, and technology in agriculture on production, employment, and income distribution.

# CONCLUSION

# CONCLUSION

> Social structures, types, and attitudes are coins that do not readily melt. Once they are formed they persist, possibly for centuries, and since different structures and types display different degrees of this ability to survive, we almost always find that actual group and national behavior more or less departs from what we should expect it to be if we tried to infer it from the dominant forms of the productive process.[1]
>
> Joseph A. Schumpeter

The factors that determine the output and productivity of land and labor are the same everywhere—sun, soil, seed, water, intensity of cropping, and husbandry. Differences in natural endowments, accidents of history, and random or deliberate acts of human judgment, however, have produced a vast array of different techniques, traditions, and institutional arrangements—farming systems—for using land and labor in different parts of the globe.

The types and implications of various policy choices of some of the developing countries of Asia and Africa have been analyzed in the text. Given the state of knowledge of the social sciences and technical data, the vast differences in the bio-physical and institutional environments in which peasants operate preclude a common prescription for improving land and labor productivity in agriculture. There can be no single policy or strategy for regulating the allocative behavior of millions of individual farm households in such different cultures, climes, and cropping regions. But it is possible to enunciate the following general principles and observations as guidelines in the formulation of policies and programs relating to the use of land and labor in agriculture in low-income countries:

**I.** In almost every developing country, except China perhaps, it is technically feasible and socially desirable to improve crop yields by a substantial increase in labor inputs per hectare/year. This would include countries in Sub-Saharan Africa with relatively favorable man-land ratios in view of: (a) the rapid rates of growth of rural population; (b) the low level of management and technical skills of the *average* cultivator; (c) the high cost of modern technology; and (d) the lack of adequate infrastructure, physical and administrative, to assure efficient delivery and use of chemical inputs or maintenance of machines and equipment in the rural areas.

**II.** The key to a successful and smooth transition from a stagnant and traditional to a dynamic and productive agriculture is not necessarily "modern" or "scientific" technology, but the *timing* and *sequencing* of technological change in cropping and cultural practices. This does not deny the need for change. Nor does it imply fixed or preordained stages of development and growth in the Marxist or Rostovian sense. On the contrary, *the order, nature, and sequencing of change would be unique to each situation and time*. But the probability of success will be far greater if new techniques (and institutions) are appropriate in relation to not only the factor endowments, natural and other resources, but also to the customary cultural practices, work patterns, and institutions governing the use of land and labor at the level of the individual and the community.

It is manifestly absurd, for instance, to introduce totally new inputs like chemical fertilizers and pesticides to illiterate peasants first, and *then* start allocating resources for investment in human capital.

It obviously takes time and a lot of resources to educate millions of adults. As in any other profession, it takes additional time for farmers to acquire the requisite experience, sophistication, and managerial competence to use new production techniques efficiently. And it takes even longer for related institutions to change or for new institutions to be established. Prices, subsidies, and profits cannot possibly eliminate the time lag. Not surprisingly, therefore, in most developing countries, whereas literacy programs are lagging grievously for lack of resources, immense amounts of money spent on the purchase of modern manufactured equipment and inputs are being largely wasted because the majority of the cultivators cannot read the manuals or instructions on the labels. And there are not, and will never be, a sufficient number of extension agents to do it for them.

On the other hand, history has repeatedly demonstrated that very simple innovations in methods of land use and cropping systems can produce dramatic results in terms of increases of production and employment.

As, for example, introduction of the turnip triggered a landmark technological revolution in English agriculture in the eighteenth century. This enabled the classic four-course rotation and integration of crop with livestock production which became the model for the rest of Western Europe.

The key features of the "revolution" were limited to changes in the organization and pattern of land use; introduction of new forage and green manure crops; and greater use of animal manures. The yield-increasing inputs were supplied entirely by the agricultural sector. But there was a substantial increase in the total output; in per hectare yields of foodgrains; and in the labor input per agricultural worker.

Introduction of the early maturing Champa rice had a similar and revolutionary impact on rice production, cropping systems, and land utilization in China. Instead of being confined to the deltas, basins, and valleys of the Yangtze River, by the thirteenth century, "much of the hilly land of the lower Yangtze region and Fukien had been turned into terraced paddies. By the close of the 16th century, Champa rice had made double, and sometimes triple, rice cropping common."[2]

Japan has demonstrated the feasibility and the immense advantages of "transforming traditionally" by small, incremental, inexpensive, step-by-step improvements in farming equipment and techniques, such as horse plowing with short bed plow; transplanting in straight lines; invention and introduction of the revolving weeder known as *tauchi-guruma*; selection of better seeds by using salt water, and so on.[3]

Japan has also shown that it is possible to modernize a peasant agriculture under a market and family type of farming system *without creating an indigent class of landless workers or an unwanted surplus of cultivating households*. During 1878 to 1962 agricultural production more than tripled, but the farm labor force remained virtually constant.

Between 1880 and 1970, moreover, the rates of growth of both total farm production and output per hectare were higher in the eastern island kingdom than in the four leading Western industrial economies of Britain, France, Germany, and the United States. In 1880, when no "scientific" methods, inputs, or equipment were in use, the average per hectare yield in terms of wheat units[a] was almost three times higher in Japan (2.86) than in the United States (0.981) *in 1970*. In the latter year, Japanese farmers produced 10.03 wheat units per hectare of cultivated land.[4] In 1972, the per hectare yield of cereals (wheat, rice, and coarse grains) was 3.9 tons in the United States and 5.5 tons in Japan—the highest in the world.[5]

> [Yet in] three hundred years Japan has not known what
> might be described as a *revolutionary* change in technology
> or organization of its agriculture.[6]

Again, not recognized generally is the fact that if the United States had as many people today as China does, and vice versa, China would be the "bread-basket" of the world in spite of its much smaller land area under cultivation, and a predominantly "traditional" technology. Historically and currently, in fact, agriculture in the East Asian region is the most productive in the world in terms of output per unit of land, the key problem in most of the low-income countries. It would be a national calamity and human tragedy of an unprecedented magnitude, for instance, if the average farm operator in China or India were to produce food to feed as many people as the American farmer does. That would create a massive displacement of surplus farm workers who could not possibly be relocated, retrained, or reemployed in any other sector of the economy for decades, if ever. Unlike surplus grain, milk, or cheese, tens of millions of men, women, and children cannot be stored, destroyed, or shipped to other lands and countries.

**III.** The low-income countries will face a predictable and certain financial ruin and bankruptcy, if in their anxiety to achieve a rapid rate of agricultural growth by "modern" methods, they become

a. Wheat unit=one ton of wheat.

entrapped in the "technology treadmill." It is pertinent to note that there is not one single capital-intensive mechanized agriculture in the industrial market economies of the world today in which the large or small farmers are able to produce and remain solvent without off-farm income, substantial loans, and price subsidies, which the meager budgets of poor countries can ill afford.

Thus, in 1976, the average off-farm income including government payments, constituted nearly 60 percent of the total income of American farmers. During fiscal 1982, the farmers received $12 billion in price supports. But the highest average payments go to large farms with the largest sales and highest farm income, farmers who generally receive high praise and accolades for their exemplary enterprise and efficiency. Without price supports their expenses would exceed receipts.[7]

In their pursuit of greater efficiency, however, as the enterprising farm operators computerized their accounts, flew their own aircraft, invested in bigger and more modern machines—12- to 14-row planters for corn and soybeans, high-power tractors, and combine harvesters—the farm debt tripled in a decade to a total of $194.5 billion on 1 January 1982. The value of *non-land* assets including machinery, motor vehicles, crop and livestock inventories and equipment, also climbed precipitously. In 1978, a typical rice farm in California with irrigation equipment required $330,000 worth of machinery while that of a spring wheat and potato farmer in North Dakota cost over $255,000. Farm expenses rose from $75.9 billion in 1975 to $141.5 billion in 1981.[8]

On 1 January 1981, the real assets of American farmers (including land but not financial securities) were valued at $1,050 billion. This massive investment in plant and technology in some 2.3 million farms produced a modest harvest of 311.5 million metric tons of food and feed grains.[9]

IV. As with new technology, illiterate farmers who cannot add, subtract, or multiply cannot be expected to *calculate* marginal costs, returns, and risk, within their individual allocative domain, to tune "so subtly to economic conditions that many experts fail to recognize how efficient they are."[10] To do so they would have to be endowed with special and unique instinct or talent which even highly educated academics, including economists and managers of modern banks and corporations, clearly lack. Farmers are normal people, neither less nor more stupid or intelligent than the rest of humanity. But complex cost-benefit analyses and calculations require more than just native intelligence.

In reality, the apparent *efficiency* of the "lowly" peasant in the allocation of his resources is due to an *unquestioning* conformity with the prescribed customary production practices of the community which evolved over several generations or centuries by a process of trial, error, and often pure chance, and not because of *deliberative*, or necessarily even accurate, *calculation* of factor and product prices and returns on investment.

> ...once an equilibrium is reached between technology and institutions, a pattern of the combination of various current inputs tends to be established. Even when relative prices of factors and inputs change, proportions of factors and inputs in use in the established technology are not subject to easy change in response, unless basic economic conditions change in the long-run.[11]

**V.** Commercial farmers cannot be expected to continue to produce at a loss. But prices alone are a poor and often inefficient instrument for manipulating agricultural output, especially per hectare *productivity*, in both the developing and developed agricultural economies. According to Leon Mears, for example, rice price policy in the United States "can be accepted as a practical upper limit on the degree to which risk can be minimized and knowledge about fertilizer effectiveness maximized because of effective price support policies," the hypothesis being that relatively higher prices draw yields up the direct price-yield function and to the right along the efficiency function. The low degree of price risk allegedly explains why yields of rice in America, and presumably of other crops as well, are on or very close to the potential efficiency frontier.[12] The assumption is that higher prices stimulate investment because the investment is profitable.

Yet in 1980 and 1981, for two consecutive back-to-back years the cost of producing a crop on the farm was more than what the American farmers received for growing it. Again, in 1982, farm income was the lowest in real dollars since 1933. Income margins for the major crops were either thin or negative. In fact, farmers were expected to earn less money—$19 billion—than they paid out in *interest alone*—$22 billion. Not surprisingly, there were nearly 7,000 voluntary or forced farm foreclosures in the first ten months of the fiscal year.

The parity price ratio (prices received index divided by prices paid) in early 1982 was down to 57 percent of the 1910-14 base and 77 percent of the more recent base of 1967. Nevertheless, and not for the first time in recent decades, despite a reduction in federal aid provided by commodity loans and target prices which meter direct Treasury payments, as well as a voluntary acreage reduction program, the American farmer kept on increasing his production, planting fence to fence. The 1982 grain harvest was the largest in history. Storage space was so tight with a 4.39 billion-bushels carryover from the previous year's bumper crop that about 1 billion bushels of grain had to be left outdoors, protected only by tarpaulins. *Time* (October 4, 1982) cited the example of Wayne Buck, 54, who farmed 900 acres of corn in central Iowa. He admitted that his stoicism was a little crazy. "I don't know why I keep banging my head against the wall," he said. But it could not possibly be because of high prices, profits, and a low risk environment.[13]

In fact, there may *not* be a *key* price or upper limit to subsidies on produce and inputs that would eliminate risk and maximize efficiency of production and factor use—the elusive "long-run functional relationship between the equilibrium price of foodgrain and the price of fertilizer (and other inputs)," generally regarded as "the heart of the political dilemma" in price policy formulation in the developing countries. An equally pertinent question would be whether elimination of *all* risk is a prerequisite for crop yields to approach the potential efficiency frontier.[14] History would indicate that the answer is: not necessarily so.

VI. Intensive use of land "presupposes a farm labor force that is prepared to devote long hours to careful tillage during the growing season and to composting, land leveling, and maintenance of irrigation works in the off-season."[15]

In most low-income countries, even with high population densities, however, most farmers seemingly choose to work far less and cultivate the land less intensively than the natural, economic, and technical conditions and opportunities would permit. It could be because of deep-rooted traditions and attitudes relating to land. Not only is the concept of land as private property relatively new and by no means universal yet in all cultures, but in many agrarian societies, man's attachment and feelings for a piece of the earth are more intimate, emotional, and irrational than for a woman. It is not regarded as a commodity. This could also be because of custom and social taboos relating to the use of family labor in a community and the types of farm work its members will or will not perform. Often they play a more decisive role in defining the potential limits to human effort and the organization of agricultural production, than the structure of landownership, technology, factor prices and endowments, or education.

It is true that not all institutional structures and customs are impervious to change. But as noted by Schumpeter, they do not always "melt" readily. The perennial cliché that peasants are not "tradition-bound," which every development economist feels compelled to reiterate *ad nauseam*, is not very meaningful. If the term implies an innate incapacity to think, reason, or experiment, then peasants are no more "tradition-bound" than doctors, lawyers, priests, or politicians. But neither are they, or *any* other group in *any* society and culture, wholly *tradition-free*—unaffected by long-established, and often obsolete, beliefs, preferences, habits, and ethnic and other prejudices. Consciously, unconsciously, and even subconsciously, traditions pervade every realm and aspect of their outlook, interpersonal relationships, and daily actions. To cite a few specific results of recognizing the influence of tradition on farmers' behavior:

—It would explain why the emaciated and even hungry Hindu farmer will not slaughter his dry and emaciated cow for meat, and the Masai is reluctant to cull the herd and sell his animals for profit.

—It would explain why in a densely populated country like China, impoverished peasants continued to allocate scarce arable land in the midst of their minuscule paddy fields—over 2.5 million acres in 1930—for the "graves of the fathers."[16] Graves were removed to wastelands only under state orders after the revolution.

—It would explain why in a modern, democratic, and prosperous Japan, the majority of the farm families continued to supply free labor for the repair and maintenance of their village roads and for desilting and weeding farm ditches just as they did in feudal times several centuries ago. Despite the sharp decline in the relative importance of agriculture in the total economy and an acute shortage of labor on the farms, only 0.3 percent or less of the work for the customary community operations of agricultural *Shuraku* (hamlet) in the country was performed by hired workers in 1970.[17]

—It would also explain why, contrary to the theory that peasants maximize total output rather than profit and therefore tend to employ family labor until the marginal productivity reaches zero, even the smallest farmers in South Asia hire labor—"much more than would be justified by the seasonality and peakedness of operations." In India, 83 percent of the operators of farms in the size group of 0-5 acres employed wage workers in 1970-71. For harvesting operations in Bangladesh, in the three seasons of *Aman, Boro,* and *Aus* rice in 1975-77,[a] hired labor performed 84, 78, and 88 percent, respectively, of the work on farms of 2.00-2.99 hectares. On farms of more than 3 hectares, the proportion increased to 98, 90, and 95 percent of the total labor input.[18]

The only logical *economic* explanation for the paradox would be that even the poorest peasants in Bangladesh and India placed a higher value on their labor than not only the market wage but that of farmers in Japan. It does not explain *why* that is so. But if that be so, it must similarly affect the entire range of their technical and allocative choices on the farm, including expectations of profit and marginal rate of return on cash expenditures. If they too are unduly high, the inevitable result would be deliberate underinvestment in land and purchased inputs, like chemical fertilizer—as indeed has been the case in the cultivation of the high-yielding varieties of rice in the region. For short-term policy purposes, therefore, it may be necessary to seek ways of persuading self-employed farmers in low-income countries to lower their subjective or "implicit wage" and returns on investment to more realistic levels in relation to the status and capacity of the national economy. History of agriculture in Japan during the Meiji period (1868-1912) is instructive in this regard.[19]

a. *Aus* and *Aman* are summer crops, sown in April. *Boro* rice is transplanted after mid-November and harvested before the wet season arrives.

**VII.** Finally, in a situation of *unrestricted choice*, that is, if individuals and communities are free to choose among several viable alternatives, their allocative, economic, and technical choices relating to the use of land and labor (and other inputs) in agricultural production will also vary, often significantly.[20]

In designing development strategies, policy-makers generally ignore the potential variations inherent in the farmers' exercise of free will in decision-making. In doing so, however, they fail to address and effectively deal with one of the most critical problems in macro-development planning—*of heterogeneity of behavior and response between and within rural societies stemming from differing traditions, values, and perceptions of costs and risks of exploiting available and even identical opportunities in agriculture.* The reluctance to take adequate cognizance of the phenomenon is based largely on the conventional theory and assumption that the differences are due to and can be explained by exogenous factors and constraints, primarily the environment of distorted incentives under which most farmers in most developing countries allegedly operate. Ironically, the most enthusiastic proponents of the theory also support greater freedom of individual and social choice and opportunity. But they assume that if farmers or peasants have access to land, inputs, and information, and if the economic incentives are "right," economic laws will transcend other constraints and differences and insure or induce a predictable measure of unanimity of choice—both individual and collective "rationality."

> All farmers—small, medium, and large...share a rationality that far outweighs differences in their social and ecological conditions.[21]

*The problem is that in real life there are no economic or natural laws of rational behavior; only man-made rules and regulations.*

# REFERENCES

REFERENCES

# REFERENCES

## PREFACE

1. Pierre Gourou, "The Quality of Land Use of Tropical Cultivators," in *Man's Role in Changing the Face of the Earth* (Chicago: University of Chicago Press, 1956), pp. 343, 344, 356.
2. Shigeru Ishikawa, *Economic Development in Asian Perspective* (Tokyo: Institute of Economic Research, Hitotsubashi University, 1967), pp. 77, fn. 34.
3. Evelyn Sakakida Rawski, *Agricultural Change and the Peasant Economy of South China* (Cambridge, Mass.: Harvard University Press, 1972), pp. 13-14.
4. A. Vaidyanathan and A.V. Jose, "Absorption of Human Labour in Agriculture: A Comparative Study of Some Asian Countries," in *Labour Absorption in Indian Agriculture: Some Exploratory Investigations*, ed. P.K. Bardhan, A. Vaidyanathan, Y. Alagh, G.S. Bhalla, and A. Bhaduri (Bangkok: ILO-ARTEP, 1978), Table 1, p. 168.
5. Gerald M. Meier, ed., *Leading Issues in Economic Development*, 3rd ed. (New York: Oxford University Press, 1976), pp. 563-64.
6. Ibid., see pp. 563-68, 575. Also see *Agricultural Policy: A Limiting Factor in the Development Process* (Washington: Inter-American Development Bank, March 1975), pp. 46-47.
7. *The Global 2000 Report to the President of the U.S.* (New York: Pergamon Press, 1981), vol. 1, p. 200; vol. 2, p. 100, Tables 6-14.


## ASIA—CHAPTER 1

1. *Asian Agricultural Survey* (published for the Asian Development Bank by University of Washington Press, 1969), pp. 22, 23.
2. *The Green Revolution*, Symposium on Science and Foreign Policy, Committee on Foreign Affairs, U.S. House of Representatives, December 5, 1969, pp. 28-31.


## ASIA—CHAPTER 2

1. *Farm-Level Constraints to High Yields in Asia: 1974-77* (Los Banos, Philippines: International Rice Research Institute, 1979), p. 45.
2. Ibid., p. 20.
3. *World Development Report, 1978* (Washington: World Bank), p. 39.
4. *Asian Agricultural Survey, 1969*, p. 557.
5. *Changes in Rice Farming in Selected Areas of Asia, Interpretative Analysis of Selected Papers* (Los Banos, Philippines: International Rice Research Institute, 1978), pp. 50, 53.

6. A.A.M. Ekramul Ahsan, "Comments on Exploring the Gap Between Potential and Actual Rice Yields: The Philippines Case," in *The Economic Consequences of the New Rice Technology* (Los Banos, Philippines: International Rice Research Institute, 1978), p. 29.

7. D.B. Grigg, *The Agricultural Systems of the World* (New York: Cambridge University Press, 1978), p. 54.

8. F.H. King, *Farmers of Forty Centuries* (Pennsylvania: Rodale Press, 1911), p. 211.

9. Ishikawa, *Economic Development in Asian Perspective*, p. 162.

10. Ibid., Appendix Table 2A-1, p. 213.

11. Kusum Nair, *The Lonely Furrow, Farming in the United States, Japan and India* (Ann Arbor: University of Michigan Press, 1969), p. 169.

12. Shigeru Ishikawa, *Labour Absorption in Asian Agriculture* (Bangkok: An ILO-ARTEP Publication, 1978), p. 29.

13. *Farm Level Constraints to High Yields*, p. 355.

14. Robert F. Chandler, Jr., *Rice in the Tropics: A Guide to the Development of National Programs* (Boulder, Colo.: Westview Press, 1979), pp. 154, 155.

15. Ping-ti-Ho, "Early-Ripening Rice in Chinese History," *Indian Journal of Agricultural Economics*, vol. 24 (Oct.-Dec. 1969), p. 207.

16. Ibid., p. 201.

ASIA—CHAPTER 3

1. John R. Freeman, "Flood Problems in China," in *Transactions of the American Society of Civil Engineers*, vol. 85 (1922), p. 1405.

2. Kang Chao, *Agricultural Production in Communist China, 1949-1965* (Madison: University of Wisconsin Press, 1970), p. 88.

3. *Sixth Five Year Plan, 1980-85* (Planning Commission, Government of India), pp. 41, 140.

4. Thomas B. Wiens, "Technological Change," in *The Chinese Agricultural Economy*, ed. Randolph Barker and Radha Sinha (Boulder, Colo.: Westview Press, 1982), p. 99.

5. *Sixth Five Year Plan* (India), p. 265.

6. Anthony M. Tang and Bruce Stone, *Food Production in the People's Republic of China* (International Food Policy Research Institute, May 1980), Table 17, p. 61.

7. Ibid., p. 137.

8. *Sixth Five Year Plan* (India), p. 104.

9. *Report of the Working Group on Energy Policy* (Planning Commission, Government of India, 1979), p. 91.

10. Ibid., p. 104.

11. *The First Five Year Plan* (Planning Commission, Government of India), p. 259.

12. *Review of the First Five Year Plan* (Planning Commission, Government of India, May 1957), p. 93.

13. *Agricultural Situation in India* (Ministry of Food and Agriculture), vol. 15, no. 5, p. 460.

14. Chao, *Agricultural Production in Communist China*, p. 148.

15. Nair, *The Lonely Furrow*, p. 209.

16. B.G. Verghese, "An Agenda for India," *Hindustan Times* (Delhi), January 18, 1970.
17. "Agenda Note on Massive Production and Utilisation of Organic Manures," mimeographed (May 2, 1973).
18. S.R. Barooah, "Efficient Use of Inputs," mimeographed (Ministry of Food and Agriculture, India, August 25, 1973).
19. "Agenda Note on Production of Organic Manures," p. 3.
20. *Sixth Five Year Plan* (India), p. 104.
21. *Labour Absorption in Indian Agriculture*, p. 204.
22. M.L. Dantwala, "The Poor Should Become Producers," in *Commerce* (Bombay), "Agriculture in the 80s" (Annual no. 1979), p. 11.
23. *Report of the Congress Agrarian Committee* (New Delhi, 1949), p. 20.
24. See *Labour Absorption in Indian Agriculture*, Table 1, p. 42; Table 14, p. 74; p. 76.
25. Thomas G. Rawski, *Economic Growth and Employment in China* (Published for the World Bank, Oxford University Press, 1979), p. 93.
26. Ibid., p. 116.
27. Chao, *Agricultural Production in Communist China*, p. 275.
28. Anthony M. Tang, "Input-Output Relations in the Agriculture of Communist China, 1952-1956," in *Agrarian Policies and Problems in Communist and Non-Communist Countries*, ed. Douglas Jackson (Seattle: University of Washington Press, 1971), p. 300; Table 3, p. 289.

## ASIA—CHAPTER 4

1. *Royal Commission on Agriculture in India*, *Report* (Bombay, 1928), p. 83.
2. "Agenda Note on Production of Organic Manures," p. 2.
3. John Augustus Voelcker, *Report on the Improvement of Indian Agriculture* (London, 1897), p. 123.
4. Ibid., pp. 118, 123, 124-25, 128.
5. *Royal Commission on Agriculture*, p. 83.
6. Ibid., pp. 75, 76, 80, 83.
7. Cited in Rawski, *Economic Growth and Employment in China*, p. 73.
8. John Lossing Buck, *Land Utilization in China* (New York: Council on Economic and Cultural Affairs, 1956), Table 14, p. 261; Table 14, p. 259; pp. 123, 260.
9. *Famine Inquiry Commission*, Final Report (New Delhi, 1945), p. 145.
10. Cited in Ping-ti-Ho, *Studies on the Population of China, 1368-1953* (Cambridge, Mass.: Harvard University Press, 1967), p. 274.
11. John Lossing Buck, *Chinese Farm Economy* (Chicago: University of Chicago Press, 1930), p. 402.
12. Ibid., pp. 402-3.
13. Voelcker, *Report on the Improvement of Indian Agriculture*, p. 101.
14. Ibid., p. 131.
15. Ibid., p. 124.
16. Dharma Kumar, *Land and Caste in India* (Cambridge: Cambridge University Press, 1965), pp. 26, 180, 190.

17. W.H. Wiser, *The Hindu Jajmani System* (Lucknow: Lucknow Publishing House, 1958), pp. 34-35.
18. *Sixth Five Year Plan* (India), pp. 408, 417, 420.
19. Voelcker, *Report on the Improvement of Indian Agriculture*, pp. 100-01, 102.
20. Kusum Nair, *Blossoms in the Dust*, Midway Reprint (Chicago: University of Chicago Press, 1979), p. 177.

ASIA—CHAPTER 5

1. Thomas A. Lumpkin, "China's Organic Fertilizers—A Threatened Tradition," Paper prepared for Workshop on Agricultural and Rural Development in China Today: Implications for the 1980s, mimeographed (Ithaca, N.Y.: Cornell University, April 1981), p. 3.
2. *The Global 2000*, vol. 1, p. 200.
3. Chao, *Agricultural Production in Communist China*, pp. 56, 89-90.
4. See Leslie T.C. Kuo, "Technical Transformation of Agriculture in Communist China," in *Agrarian Policies and Problems in Communist and Non-Communist Countries*, pp. 250-74. Also, Sen-dou-Chang, "Restructuring Agricultural Landscape in China," in *The China Geographer*, no. 11, ed. C.W. Pannell and C.L. Salter (Boulder, Colo.: Westview Press, 1981), p. 18.
5. *A Survey of the Fertilizer Sector in India*, World Bank Staff Working Paper no. 331 (Washington, June 1979), pp. 32, 33, A-54.
6. *Industrialization, Technology and Employment in the People's Republic of China*, World Bank Staff Working Paper no. 291 (Washington, August 1978), p. 74.
7. Wiens, in *The Chinese Agricultural Economy*, pp. 110-11, 115-17.
8. Rawski, *Economic Growth and Employment in China*, see Ch. 5.
9. Ibid., pp. 116, 117, 131.
10. Ibid., pp. 98-100.
11. Ibid., pp. 95-96.
12. Ibid., p. 110.
13. James E. Nickum, ed., *Water Management Organization in the People's Republic of China* (New York: M.E. Sharpe, 1981), p. 32.
14. Rawski, *Economic Growth and Employment in China*, p. 113.
15. Ibid., pp. 128-29.
16. Dantwala, in *Commerce*, 1979, p. 9; S.L. Shah, "Technological Innovation and Growth in Indian Agriculture," *Commerce*, 1979, p. 59.
17. *The Sixth Five Year Plan* (India), p. 305.
18. Ibid., pp. 51, 170.
19. Dantwala, in *Commerce*, 1979, p. 9.
20. John W. Mellor, *The New Economic Growth* (Ithaca, N.Y.: Cornell University Press, 1976), p. 164. Also see Rawski, *Economic Growth and Employment in China*, pp. 145-46.

AFRICA—CHAPTER 6

1. *Food Problems and Prospects in Sub-Saharan Africa: The Decade of the 1980's*, Foreign Agricultural Research Report no. 166 (Washington: U.S. Department of Agriculture, August 1981), pp. 87-88.

## AFRICA—CHAPTER 7

1. *Accelerated Development in Sub-Saharan Africa: An Agenda for Action* (Washington: World Bank, 1981), p. 2.
2. Ibid., p. 3.
3. Uma Lele, "Rural Africa: Modernization, Equity, and Long-Term Development," *Science*, vol. 211, February 6, 1981, p. 550.
4. *Accelerated Development in Sub-Saharan Africa*, p. 77.
5. *Agricultural Development Indicators* (New York: International Agricultural Development Service, 1981), pp. 13, 15.

## AFRICA—CHAPTER 8

1. Kenneth R.M. Anthony, Bruce F. Johnston, William O. Jones, Victor C. Uchendu, *Agricultural Change in Tropical Africa* (Ithaca, N.Y.: Cornell University Press, 1979), p. 102.
2. *Accelerated Development in Sub-Saharan Africa*, p. 48.
3. *Agricultural Change in Tropical Africa*, p. 115.
4. Derek Byerlee, Carl K. Eicher, Carl Liedholm, and Dunstan S.C. Spencer, *Rural Employment in Tropical Africa: Summary of Findings*, Working Paper no. 20 (East Lansing: Department of Agricultural Economics, Michigan State University, February 1977), p. 14.
5. Ibid., p. 21.
6. *Agricultural Change in Tropical Africa*, p. 40, fn. 11.
7. Ibid., pp. 92-93; Tables 3-8, p. 94.
8. *Rural Employment in Tropical Africa*, p. 15.
9. Ibid., p. 167.
10. *Food Problems and Prospects in Sub-Saharan Africa*, p. 71.
11. *Agricultural Change in Tropical Africa*, p. 18.

## AFRICA—CHAPTER 9

1. *Accelerated Development in Sub-Saharan Africa*, p. 24.
2. *Food Problems and Prospects in Sub-Saharan Africa*, p. 258.
3. *Accelerated Development in Sub-Saharan Africa*, p. 30.
4. Ibid., p. 58.
5. Kusum Nair, *Three Bowls of Rice; India and Japan: Century of Effort* (East Lansing: Michigan State University, 1973), pp. 97, 103.
6. Ibid., p. 103.
7. Milton and Rose Friedman, *Free to Choose* (New York: Harcourt Brace Jovanovich, 1980), pp. 190-91.
8. *Economic Report of the President, February 1982* (Washington: U.S. Govt. Print. Off., 1982), pp. 134, 136.
9. *Food Problems and Prospects in Sub-Saharan Africa*, p. 118.
10. *Accelerated Development in Sub-Saharan Africa*, p. 66, fn. 27.
11. Ibid., pp. 57, 59, 64.
12. John Kenneth Galbraith, *Money: Whence It Came, Where It Went* (Boston: Houghton Mifflin Co., 1975), p. 245.

13. Ibid., pp. 247, 249.
14. George Rogers Taylor, *The Transportation Revolution, 1815-1860*, vol. 4, *The Economic History of the United States* (New York: M.E. Sharpe, 1951), pp. 378-80.

AFRICA—CHAPTER 10

1. *Science*, February 6, 1981, p. 553.
2. *World Development Report 1981*, p. 102.
3. See Michael Lipton, "Botswana: Employment and Labor Use," vols. 1 and 2, mimeographed (Gaborone, Botswana: Published by Government Printer, 1979); and Lipton, "Botswana's Agriculture in Light of Asian Experience," IDM Lecture, May 3, 1978.
4. *Food Problems and Prospects in Sub-Saharan Africa*, p. 40.
5. Kenneth H. Parsons, "Customary Land Tenure and the Development of African Agriculture," LTC no. 77, mimeographed (Madison: Land Tenure Center, University of Wisconsin, June 1971), p. 73.
6. Ibid., pp. 26, 82.
7. Kenneth H. Parsons, "Land Policy in Development of Agriculture: Two Fragments with a Preface," mimeographed (Madison: Land Tenure Center, University of Wisconsin, March 1982), Part 2, p. 14.
8. Ibid., Part 1, p. 43.
9. David J. King, "Land Reform and Participation of the Rural Poor in the Development Process of African Countries," LTC no. 101, mimeographed (Madison: Land Tenure Center, University of Wisconsin, September 1974), p. 49.
10. Richard L. Barrows, "Individualized Land Tenure and African Agricultural Development: Alternatives for Policy," LTC no. 85, mimeographed (Madison: Land Tenure Center, University of Wisconsin, April 1973), p. 13.
11. Ibid.
12. *Kenya: Into the Second Decade*, World Bank Report (Baltimore: Johns Hopkins University Press, 1975), p. 475.
13. Michael Hechter, *Internal Colonialism* (Berkeley and Los Angeles: University of California Press, 1977), p. 86. Also see pp. 72, 102-03.
14. Parsons, "Customary Land Tenure," p. 7.
15. Ibid., pp. 4, 8.
16. Nair, *The Lonely Furrow*, p. 16.
17. Lester R. Brown, *Building a Sustainable Society* (New York: W.W. Norton & Co., 1981), p. 21.
18. Barrows, "Individualized Land Tenure," p. 11.
19. Clough and Cole, *Economic History of Europe*, rev. ed. (Boston: D.C. Heath & Co., 1947), p. 191.
20. *Kenya: Into the Second Decade*, p. 5.
21. *Poverty and Growth in Kenya*, World Bank Staff Working Paper no. 389 (Washington: World Bank, May 1980), pp. 9, 10.
22. Barrows, "Individualized Land Tenure," p. 16.
23. See *Rural Poverty: Hearings before the National Advisory Commission on Rural Poverty* (Washington, D.C., 1967), esp. pp. 47, 107-08, 446, 447, 490-92.

24. Parsons, "Land Policy in Development of Agriculture," Part 2, pp. 16-17.
25. Ibid., Part 2, pp. 29-30.
26. Ibid., Part 2, p. 41.
27. Ibid., Part 2, p. 45.

AFRICA—CHAPTER 11

1. *Food Problems and Prospects in Sub-Saharan Africa*, p. 89.
2. *Accelerated Development in Sub-Saharan Africa*, p. 51.
3. *Food Problems and Prospects in Sub-Saharan Africa*, p. 79.
4. Mellor, Ruttan, Schultz, among others.
5. D.B. Grigg, *The Agricultural Systems of the World* (New York: Cambridge University Press, 1978), p. 53.
6. John C. de Wilde et al., *The Synthesis*, vol. 1, *Experiences with Agricultural Development in Tropical Africa* (Baltimore: Johns Hopkins University Press, 1967), p. 106.
7. *Growth with Equity: An Economic Policy Statement* (Government of the Republic of Zimbabwe), p. 89.
8. de Wilde et al., *Experiences with Agricultural Development in Tropical Africa*, p. 106.
9. Ibid., p. 107, fn. 22.
10. *Agricultural Change in Tropical Africa*, p. 190.
11. *Food Problems and Prospects in Sub-Saharan Africa*, p. 100.
12. Malcolm Christie and Guy Scott, *Zambia: An Agricultural Development Strategy for the Next 25 Years*, EDI Seminar Paper no. 14 (Washington: Economic Development Institute, World Bank, March 1977), p. 7.
13. Ibid., p. 11.
14. *Food Problems and Prospects in Sub-Saharan Africa*, p. 141.
15. Ibid., pp. 158, 159.
16. Dunstan S.C. Spencer, "Private and Social Profitability in Rice Production and Marketing in Sierra Leone," Paper prepared for project funded by U.S. Agency for International Development, mimeographed (July 1979), p. 28.
17. *Food Problems and Prospects in Sub-Saharan Africa*, p. 118.
18. Julius K. Nyerere, *Ujamaa—Essays on Socialism* (New York: Oxford University Press, 1968), pp. 22, 23, 33.
19. Source: *World Bank Reports*.
20. *Accelerated Development in Sub-Saharan Africa*, pp. 3-4.
21. *Food Problems and Prospects in Sub-Saharan Africa*, p. 105.
22. *Science*, February 6, 1981, p. 552.
23. *Accelerated Development in Sub-Saharan Africa*, p. 74.
24. Peter Oghyayerio Agbonifo, "State Farms and Rural Development: A Case Study of the Agbede and Warrake Farm Projects in Bendel State of Nigeria," Ph.D. thesis, University of Wisconsin-Madison, December 1980, p. 29.
25. *Rural Employment in Tropical Africa*, p. 55.
26. J. Dirck Stryker, "Comparative Advantage and Public Policy in West African Rice," Paper prepared for Stanford University Food Research Institute project funded by USAID, mimeographed (July 1979), p. 19. See also Scott R. Pearson, Charles P. Humphreys, and Eric A. Monke, "A Comparative Analysis of Rice Policies in Five West African Countries," mimeographed, p. 13.

27. John McIntire, "Rice Policy in Mali," Paper prepared for project funded by USAID, mimeographed (July 1979), Table A-3 (Appendix).
28. *Zambia: An Agricultural Development Strategy*, pp. 3, 4-5.
29. J.I. Kirkwood, E. Brams, and Y.P. Chang, *Worldwide Rural Poverty: An Agronomist's Concern*, Research Report (Department of Agronomy, Prairie View A & M College), p. 62.
30. *Food Problems and Prospects in Sub-Saharan Africa*, p. 111.
31. Vernon C. Johnson and Harold Jones, "Supporting Food Production in Africa," AFR/DR/ARD, mimeographaed (October 14, 1980), p. 15.
32. *Food Problems and Prospects in Sub-Saharan Africa*, pp. 112-13.
33. *Accelerated Development in Sub-Saharan Africa*, p. 70.
34. *Food Problems and Prospects in Sub-Saharan Africa*, p. 112.
35. *Accelerated Development in Sub-Saharan Africa*, p. 72.

AFRICA—CHAPTER 12

1. *Agricultural Change in Tropical Africa*, p. 23.
2. Bernhard Glaeser and Kevin D. Phillips-Howard, "A Technological Alternative for Energy Use in Rural Development: The Case of Southeast Nigeria," Paper presented at Third International Conference on Energy Use Management, held in Berlin, October 1981, mimeographed, p. 3.4.
3. *Ujamaa—Essays on Socialism*, pp. 89-90.
4. *Growth with Equity: An Economic Policy Statement* (Zimbabwe), p. 39.
5. *Accelerated Development in Sub-Saharan Africa*, p. 52.
6. Frances Stewart, *International Technology Transfer: Issues and Policy Options*, World Bank Staff Working Paper no. 344 (Washington: World Bank, July 1979), p. 108.
7. *Lagos Plan of Action* (OAU Heads of State and Government, April 1980), Annex I, pp. 9, 19, 23; and CM/Res. 759 (33), p. 69.

CONCLUSION

1. Joseph A. Schumpeter, *Ten Great Economists: From Marx to Keynes* (New York: Oxford University Press, 1969), p. 14.
2. "Agriculture, History of." In *Encyclopaedia Britannica*, 15th ed. (1974), p. 329.
3. ARTEP, *Employment Expansion in Asian Agriculture: A Comparative Analysis of South Asian Countries* (Bangkok: International Labour Organisation, March 1980), pp. 48, 52.
4. Hans Binswanger, Vernon Ruttan et al., *Induced Innovation* (Baltimore: Johns Hopkins University Press, 1978), pp. 49, 83, 86, 87; Tables 3-2, 3-11, 3-12, 3-13, 3-14, 3-15, 3-16. Also see *The Lonely Furrow*, Ch. 12.
5. *The World Food Situation and Prospects to 1985*, Foreign Agricultural Report no. 98 (Washington: U.S. Department of Agriculture), p. 64, Table 30.
6. Nair, *The Lonely Furrow*, p. 176.
7. *Changing Character and Structure of American Agriculture: An Overview* (Washington: U.S. General Accounting Office, 1978), pp. 50, 88, 89, 121, 128.
8. Ibid., pp. 19, 26, 27, 39, 40; and USDA reports.

9. In 1981-82. The grains include wheat, oats, corn, barley, and rice. Source: Washington: Foreign Agricultural Service, USDA.

10. Theodore W. Schultz, *Investing in People* (Berkeley: University of California Press, 1981), p. 8. Also, Schultz, *Transforming Traditional Agriculture* (New Haven, Conn.: Yale University Press, 1964), Ch. 3.

11. Ishikawa, in *Employment Expansion in Asian Agriculture*, p. 45.

12. See Leon A. Mears and Peter Timmer, *Food Research Institute Studies*, vol. 14, no. 4 (1975), pp. 319-57, 419-31.

13. Data collected from a variety of sources, including: Harold F. Breimyer, "Farm Depression," *Challenge*, July/Aug. 1982; *U.S. News and World Report*, Nov. 2, 1981, and Aug. 23, 1982; *Time*, Oct. 4, 1982; *New York Times*, March 28, 1982.

14. Peter Timmer, "The Political Economy of Rice in Asia: Lessons and Implications," *Food Research Institute Studies*, vol. 14, no. 4 (1975), pp. 422, 431.

15. Rawski, *Economic Growth and Employment in China*, pp. 141-42.

16. King, *Farmers of Forty Centuries*, see Ch. 2. Also Buck, *Land Utilization in China*, pp. 176-79.

17. Ishikawa, in *Employment Expansion in Asian Agriculture*, pp. 40, 46, 54.

18. I.J. Singh, "Small Farmers and the Landless in South Asia," mimeographed (World Bank, October 1981), pp. III. 45-48.

19. See *The Lonely Furrow*, Ch. 12, 20, 21, 22.

20. For several examples of differential response, see author's *Blossoms in the Dust, The Lonely Furrow*, and *In Defense of the Irrational Peasant* (Chicago: University of Chicago Press, 1979).

21. *World Development Report 1982*, pp. 5-6.

# APPENDIX—INTERNATIONAL STATISTICS

## TABLE 1

Regional Distribution of Land, Cropland, Agricultural Population and Area per Person in Agriculture

| REGION | LAND AREA (million hectares) | CROPLAND (million hectares) | CROPLAND Distribution (%) | RURAL POPULATION (millions) | RURAL POPULATION Distribution (%) | AGRICULTURAL POPULATION AS PERCENTAGE OF TOTAL POPULATION | CROPLAND AREA PER RURAL PERSON (hectares) |
|---|---|---|---|---|---|---|---|
| Europe | 493 | 145 | 10.0 | 89 | 4.8 | 17 | 1.63 |
| USSR | 2,240 | 232 | 15.9 | 77 | 4.2 | 32 | 3.01 |
| North and Central America | 2,242 | 271 | 18.6 | 54 | 2.9 | 17 | 5.02 |
| South America | 1,783 | 84 | 5.8 | 74 | 4.0 | 39 | 1.14 |
| Asia | 2,753 | 463 | 31.8 | 1,314 | 71.0 | 64 | 0.35 |
| Africa | 3,031 | 214 | 14.7 | 239 | 12.9 | 67 | 0.90 |
| Oceania | 851 | 47 | 3.2 | 4 | 0.2 | 4 | 11.75 |
| Total | 13,393 | 1,456 | 100.0 | 1,851 | 100.0 | 51 | 0.78 |

SOURCE: FAO, *Production Year Book 1972.*

TABLE 2

Rural Population and Rural Poverty in Developing Countries

| REGION | RURAL POPULATION 1969 | RURAL POPULATION IN POVERTY | | | PERCENTAGE OF RURAL POOR IN RURAL POPULATION | | |
|---|---|---|---|---|---|---|---|
| | | Population with Incomes below $50 per Capita | Population with Incomes below $75 per Capita (millions) | Population with Incomes below 1/3 of National Average per Capita Income, or below $50 per Capita | Population with Incomes below $50 per Capita | Population with Incomes below $75 per Capita | Population with Incomes below 1/3 of National Average per Capita Income, or below $50 per Capita (percentage) |
| Developing countries: | | | | | | | |
| in Africa | 280 | 105 | 140 | 115 | 38 | 50 | 41 |
| in America | 120 | 20 | 30 | 45 | 17 | 25 | 38 |
| in Asia | 855 | 355 | 525 | 370 | 42 | 61 | 43 |
| Developing countries total | 1,255 | 480 | 695 | 530 | 38 | 55 | 42 |
| Four Asian countries[a] | 625 | 295 | 435 | 295 | 47 | 70 | 47 |
| Other countries | 630 | 185 | 260 | 235 | 29 | 41 | 37 |
| Share of developing countries | | | (percentages) | | | | |
| in Africa | 22 | 22 | 20 | 22 | | | |
| in America | 10 | 4 | 4 | 8 | | | |
| in Asia | 68 | 74 | 76 | 70 | | | |
| Total share of four Asian countries[a] | 50 | 62 | 63 | 56 | | | |

SOURCE: World Bank, *The Assault on World Poverty* (Baltimore: Johns Hopkins University Press, 1975), Annex 3, p. 80.
[a]Bangladesh, India, Indonesia, and Pakistan.

# APPENDIX—ASIAN STATISTICS

## TABLE 3

Rainfall Characteristics, Irrigation, and Cropping Intensity in Selected Asian Countries

| | MEAN TEMPERATURE IN THE HOTTEST MONTH (°C) | AVERAGE ANNUAL RAINFALL (mm) | PROPORTION OF TOTAL RAINFALL IN FOUR RAINIEST MONTHS (%) | PROPORTION OF CROPPED AREA UNDER IRRIGATION (%) | CROPPING INTENSITY |
|---|---|---|---|---|---|
| South Korea (Seoul) | 25.4 | 1259 | 71 | 26 | 152 |
| Japan (Hokkaido and Tokyo) | 17.5-26.4 | 994-2337 | 46-56 | 46 | 116 |
| China: | | | | 51 | 128 |
| Spring wheat and winter wheat millets region | 26.4 | 343-508 | 66-85 | 15 | 114 |
| Winter wheat Kaoling region | 25 | 633 | 80-85 | 28 | 154 |
| Szechwan rice region | 27.2 | 803-1105 | 54-75 | 48 | 171 |
| Yangtse rice wheat region | 28.3 | 1148-1593 | 49-56 | 56 | n.a. |
| South-west rice region | 23.6 | 1044-1176 | 57-71 | 28 | 187 |
| Double crop rice region | 28.6 | 1486-1615 | 51-59 | 52 | 119 |
| Bangladesh | 29.3 | 1880 | 66 | 5 | 181 |
| Taiwan (Taipei) | 28.4 | 2100 | 51-59 | 51 | 96 |
| Thailand (Bangkok) | 30.2 | 1252 | 61 | 22 | 102 |
| Philippines (Manila) | 28.1 | 2121 | 70 | 9 | 95 |
| Sri Lanka (Colombo) | 28.0 | 2397 | 40 | 22 | 99 |
| India: | | | | 21 | |
| North | 34.3 | 310-1214 | 78-91 | 28 | 111 |
| East | 31.1 | 1214-2380 | 65-85 | 23 | 106 |
| West | 33.5 | 686-3335 | 52-94 | 8 | 96 |
| South | 32.7 | 660-2722 | 27-81 | 30 | 95 |

(Contd.)

TABLE 3 *(Contd.)*

| | MEAN TEMPERATURE IN THE HOTTEST MONTH (°C) | AVERAGE ANNUAL RAINFALL (mm) | PROPORTION OF TOTAL RAINFALL IN FOUR RAINIEST MONTHS (%) | PROPORTION OF CROPPED AREA UNDER IRRIGATION (%) | CROPPING INTENSITY |
|---|---|---|---|---|---|
| W. Malaysia (Penang) | 28.1 | 2647 | 51 | 13 | 102 |
| Pakistan (Karachi) | 30.4 | 204 | 82 | 72 | 87 |

SOURCE: ILO-ARTEP, *Labour Absorption in Indian Agriculture*, Table 2, pp. 173-74.

1. Temperature and rainfall data relating to most of the centers (unless otherwise specified) are taken from World Climatic Table in H.H. Lamb, *Climate, Present, Past and Future*, vol. I (Methuen and Co., 1972), pp. 533-40. Rainfall data for Bangladesh (Dacca), Thailand (Bangkok), and W. Malaysia (Kuala Lumpur) are from R.R. Rawson, *The Monsoon Lands of Asia*, Table 1 (Hutchinson Educational, 1963). Temperature data for Bangladesh (Dacca). W. Malaysia (Kuala Lumpur), Thailand (Bangkok), N. India (Delhi), E. India (Calcutta), W. India (Ahmedabad), and S. India (Madras) are taken from Victor Showers, *The World in Figures* (John Wiley and Sons, 1973), pp. 192 and 348.

2. The figures relating to cropping intensity (i.e., gross cropped area divided by area of arable land) and the percentage of area irrigated for S. Korea, Taiwan, Thailand, Philippines, and W. Malaysia have been taken from IWP, vol. 2 (FAO). The figures for Japan are computed from the data published in the FAO *Production Year Book* (1975). However, this is much lower than Ishikawa's figure (1.33) based on the Farm Economic Survey data. The discrepancy needs to be explained. For China, the irrigation ratio refers to the percentage of cultivated area under irrigation, and has been obtained from Kang Chao, *Agricultural Production in Communist China* (University of Wisconsin Press, 1970), pp. 289-95. The cropping intensity figures for various regions of China are taken from S. Ishikawa, "Changes in the Structure of Agricultural Production in Mainland China," in *Agrarian Policies and Problems in Communist and Non-Communist Countries*, ed. W.A. Douglas Jackson (Seattle: University of Washington Press, 1971). The regions in Ishikawa's classification are not strictly comparable to those included in the table. The irrigation ratio and cropping intensities for different parts of India have been computed from data for 1964-65 published in *Indian Agriculture in Brief*, 11th ed. (Government of India, Ministry of Agriculture, 1971). The cropping intensity has been estimated by dividing gross cropped area by the arable land area (which includes net sown area and also the fallow lands). For Bangladesh and Pakistan the figures correspond to the average for 1961-65 and are from *Production Year Book*, vol. 28.1 (FAO, 1971).

## TABLE 4

International Comparison of National Income in
Selected Countries

| | AVERAGE ANNUAL GROWTH AT CONSTANT PRICES (%) | INCOME LEVEL (1979 $) |
|---|---|---|
| China (1957-79) | 3.5 (2.7)[a] | 256 |
| India (1960-78) | 1.4 | 190 |
| Indonesia (1960-78) | 4.1 | 380 |
| Sri Lanka (1960-78) | 2.0 | 230 |
| Low-income countries (1960-78) | 1.6 | 230 |

SOURCE: Various sources, including the *World Bank Atlas*, 1980.

[a] Figure in brackets is adjusted for relative price difference.

## TABLE 5

Cultivated Area Per Capita of the Agricultural
Population, 1978, in Selected Countries

| | AREA (ha) |
|---|---|
| Japan | 0.25 |
| Netherlands | 0.78 |
| Egypt | 0.15 |
| Republic of Korea | 0.14 |
| Indonesia | 0.16 |
| Bangladesh | 0.15 |
| India | 0.42 |
| China | 0.12 |

SOURCE: FAO, *Production Year Book*, vol. 33 (1979). For China, from various sources.

## TABLE 6

Number of Agricultural Workers, Arable Land, and Productivity of Arable Land in Selected Asian Countries

| | NUMBER OF AGRICULTURAL WORKERS (000) | | | ARABLE LAND (000 ha) | NUMBER OF WORKERS PER 1000 HECTARES OF ARABLE LAND | GROSS VALUE OF CROP AND LIVESTOCK PRODUCTION PER HECTARE OF ARABLE LAND (US $) |
|---|---|---|---|---|---|---|
| | Male | Female | Total | | | |
| South Korea | 3382 | 1765 | 5147 | 2100 | 2451 | 791 |
| Japan | 6120 | 6610 | 12730 | 5893 | 2160 | 617 |
| China | n.a. | n.a. | 236907 | 118940 | 1992 | 227 |
| Bangladesh | 13479 | 2596 | 16075 | 8919 | 1802 | 242 |
| Taiwan | 1161 | 307 | 1468 | 890 | 1649 | 663 |
| Thailand | 5580 | 5762 | 11342 | 8600 | 1319 | 124 |
| Philippines | 6657 | 1158 | 7815 | 7890 | 990 | 126 |
| Sri Lanka | 1390 | 457 | 1847 | 1870 | 988 | 258 |
| India | 88580 | 48988 | 137568 | 161500 | 852 | 99 |
| W. Malaysia | 843 | 401 | 1245 | 2350 | 530 | 248 |
| Pakistan | 7892 | 915 | 8787 | 17874 | 492 | 80 |

SOURCE ILO-ARTEP, *Labour Absorption in Indian Agriculture*, Table 1, pp. 168-69.

1. Number of workers in agriculture relates to 1960 and the data for all countries except Japan and W. Malaysia are taken from *Production Year Book*, vol. 28.1 (FAO, 1974). Taiwanese data are for 1956 and were obtained from FAO *Production Year Book* (1965). These totals have been broken down by sex on the basis of the proportion of male and female workers in agriculture as reported in the *Year Book of Labour Statistics* (ILO, 1971 and 1973). Data for W. Malaysia are from *ILO Year Book* and refer to 1957.

2. In general the figures cover those engaged in crop husbandry, livestock, fisheries, forestry and hunting (=ISIC Div. 0). However, in the case of few countries, where fisheries are more prominent, the figures exclude estimated employment in fisheries. The estimate for the agricultural work force in Japan, excluding those in the fisheries sectors, was directly obtained from *OECD Labour Force Statistics*

(Paris, 1973). For South Korea and Taiwan, the number of workers in fisheries was estimated from data given in the *"Survey of Asian Agriculture"* (Asian Development Bank, 1971) and deducted from the number of total agricultural workers. As for the other countries, the proportion of fishermen was relatively low and hence no adjustments were made.

3. Arable land includes area under seasonal crops, permanent crops and current fallows. Data for countries other than Japan, China, Pakistan and Bangladesh have been obtained from *Indicative World Plan for Agricultural Development to 1975 and 1985, Asia and the Far East* (IWP), vol. 2 (Rome: FAO, 1968). The figures correspond to the average of the three years 1961 to 1963. For Japan, China, Pakistan and Bangladesh, they have been collected from *Production Year Book, 1975*, vol. 29 (FAO/Rome), and give the average for the five years 1961 to 1965.

4. The value of crop and livestock output used in deriving the averages shown in column 6 are taken from the *Indicative World Plan* (IWP), vol. 2, pp. 24-32, in respect of the following countries: South Korea, Taiwan, Thailand, Philippines, Sri Lanka, India and W. Malaysia. All the figures relate to 1961-1963.

For the remaining four countries, they have been estimated by applying unit values of various crops and livestock products for Asia, implied in the IWP estimates, to average the production figures for 1961-65 taken from the FAO *Production Year Book*, vol. 28.1 (1974).

## TABLE 7

Estimates of the Proportion of Rice Area in Five Major Environmental Categories

| | TOTAL RICE AREA[a] (000 ha) | PROPORTION OF AREA (percent) | | | | Second Crop |
|---|---|---|---|---|---|---|
| | | Irrigated | Rainfed | Upland | Deep-water | |
| Bangladesh | 9,766 | 16 | 39 | 19 | 26 | 10 |
| Burma | 4,985 | 17 | 81 | 1 | 1 | 1 |
| India | 37,755 | 40 | 50 | 5 | 5 | 5 |
| Indonesia | 8,482 | 47 | 31 | 17 | 5 | 19 |
| Malaysia (West) | 771 | 77 | 20 | 3 | 0 | 50 |
| Nepal | 1,200 | 16 | 76 | 9 | 0 | 0 |
| Pakistan | 1,518 | 100 | 0 | 0 | 0 | 0 |
| Philippines | 3,488 | 41 | 48 | 11 | 0 | 14 |
| Sri Lanka | 604 | 61 | 37 | 2 | 0 | 25 |
| Thailand | 7,037 | 11 | 80 | 2 | 7 | 2 |
| Vietnam | 2,713 | 15 | 60 | 5 | 20 | 5 |
| Total | 78,319 | 19 | 47 | 10 | 10 | 14 |

SOURCE: R. Barker, H.E. Kauffman, and R.W. Herdt, "Production Constraints and Priorities for Research," mimeographed (Los Banos: International Rice Research Institute, April 1975).

[a] 1970-74 average area. FAO data.

### TABLE 8

Areas Harvested and Average Yields of Paddy, 1963-67 and 1973-77, and Paddy Production, 1971-75, 1976, and 1977, in Selected DMCs

| DMCs | HARVESTED AREA 1963-67 (000 ha) | HARVESTED AREA 1973-77 (000 ha) | GROWTH RATE (% p.a.) | YIELD 1963-67 (tons/ha) | YIELD 1973-77 (tons/ha) | GROWTH RATE (% p.a.) | PRODUCTION 1971-75 (000 tons) | PRODUCTION 1976 (000 tons) | PRODUCTION 1977 (000 tons) | INCREASE 1976-77 (%) |
|---|---|---|---|---|---|---|---|---|---|---|
| **SOUTH ASIA** | | | | | | | | | | |
| Afghanistan | 215 | 212 | 0.2 | 1.68 | 2.09 | 2.2 | 396 | 457 | 460 | + 0.7 |
| Bangladesh | 9,311 | 10,120 | 0.8 | 1.69 | 1.78 | 0.5 | 16,810 | 17,644 | 19,602 | +11.1 |
| India | 35,386 | 38,751 | 0.8 | 1.46 | 1.74 | 1.8 | 64,469 | 64,245 | 74,234 | +15.5 |
| Nepal | 1,100 | 1,245 | 1.2 | 1.93 | 2.01 | 0.4 | 2,433 | 2,404 | 2,670 | +11.1 |
| Pakistan | 1,373 | 1,653 | 1.9 | 1.47 | 2.35 | 4.8 | 3,573 | 4,110 | 4,272 | +3.9 |
| Sri Lanka | 517 | 629 | 2.1 | 1.94 | 2.23 | 1.4 | 1,355 | 1,253 | 1,707 | +36.2 |
| All | 48,402 | 52,610 | 0.8 | 1.52 | 1.78 | 1.6 | 89,036 | 90,113 | 102,945 | +14.2 |
| **SOUTHEAST ASIA** | | | | | | | | | | |
| Burma | 4,785 | 5,038 | 0.5 | 1.62 | 1.79 | 1.0 | 8,384 | 9,312 | 8,799 | -5.5 |
| Cambodia | 2,231 | 1,063 | -6.9 | 1.14 | 1.28 | 2.1 | 1,451 | 1,800 | 1,600 | -11.1 |
| Indonesia | 7,249 | 8,479 | 1.6 | 1.76 | 2.67 | 4.3 | 21,169 | 23,301 | 22,794 | -2.2 |
| Lao PDR | 364 | 680 | -2.3 | 0.82 | 1.25 | 4.3 | 865 | 850 | 800 | -5.9 |
| Malaysia | 552 | 749 | 3.1 | 2.16 | 2.61 | 1.9 | 1,675[a] | 1,748[a] | 1,722 | -1.5 |
| Philippines | 3,159 | 3,555 | 1.2 | 1.30 | 1.74 | 3.0 | 5,453 | 6,456 | 6,890 | +6.7 |
| Thailand | 6,402 | 8,016 | 2.3 | 1.86 | 1.83 | -0.2 | 14,003 | 15,800 | 15,000 | -5.1 |
| Viet Nam SRO | | | | | | | 7,057 | 11,782 | 11,256 | -4.5 |
| All (reporting) | 25,442 | 27,580 | 0.8 | 1.61 | 2.05 | 2.4 | 60,057 | 71,049 | 68,861 | -3.1 |
| **EAST ASIA** | | | | | | | | | | |
| China, Rep. of | 772 | | | 3.90 | 4.08[b] | 0.6 | 3,199 | 3,560 | 3,542 | -0.5 |
| Korea, Rep. of | 1,209 | 1,208 | -0.1 | 4.26 | 5.53 | 2.6 | 5,912 | 7,243 | 8,342 | +15.2 |
| All | 1,981 | | | 4.12 | | | 9,111 | 10,803 | 11,884 | +10.0 |

SOURCE: FAO, December 1977, for area and yield; USDA, *Foreign Agriculture Circular*, May 17, 1978, for production. *Sector Paper on Agriculture and Rural Development*, ADB Staff Working Paper (April 1979). Tables 1-8, Appendix p. 97.

[a] Peninsular Malaysia only.
[b] 1971-75 (ADB, *Key Indicators*, April 1978).

## TABLE 9

### Fertilizer Use in Selected DMCs, 1950, 1960, 1970, 1973

| COUNTRY | KG NPK/HA ARABLE LAND | | | |
| --- | --- | --- | --- | --- |
| | 1950/51 | 1960/61 | 1970/71 | 1973/74 |
| SOUTH ASIA | 0.7 | 2.3 | 13.7 | 17.9 |
| Bangladesh | 0.1 | 2.6 | 15.7 | 19.4 |
| India | 0.6 | 1.8 | 13.1 | 17.2 |
| Nepal | — | 0.03 | 2.7 | 7.1 |
| Pakistan | 0.3 | 4.3 | 14.7 | 20.8 |
| Sri Lanka | 18.6 | 39.1 | 47.3 | 48.0 |
| | | | | |
| SOUTHEAST ASIA | 1.0 | 3.9 | 13.5 | 18.8 |
| Burma | 0.01 | 0.3 | 1.2 | 3.0 |
| Indonesia | 1.2 | 3.5 | 13.2 | 26.3 |
| Malaysia | — | 11.4 | 46.8 | 74.3 |
| Philippines | 5.2 | 15.2 | 30.6 | 34.4 |
| Thailand | 0.2 | 1.7 | 6.3 | 11.5 |
| Vietnam | 1.9 | 8.4 | 64.5 | 59.6 |
| | | | | |
| EAST ASIA | 100.4 | 256.7 | 330.0 | 377.0 |
| China, Republic of | 92.7 | 203.8 | 283.5 | 336.2 |
| Korea | 9.1 | 139.2 | 245.7 | 317.2 |

SOURCE: *Asian Agricultural Survey 1976*, Appendix I-8.5, p. 416.

    For NPK Consumption:

        1950/51-1952/53, FAO, *Production Year Book*, 1952, 1954, 1953/54-1973/74, FAO, *Annual Fertilizer Review*.

    For India:

        1952/53-1972/73, The Fertilizer Association of India, *Fertilizer Statistics*, 1971/72-1974/75.

    For Bangladesh:

        1952/53-1973/74, Ministry of Agriculture, *Bangladesh Agriculture in Statistics*.

    For China:

        1971/1973, Asian Productivity Organization, *Impact of Fertilizer Shortage: Focus on Asia*, 1974.

SOURCE: For arable land:

    FAO, *Production Year Book*, various issues.

    *Taiwan Agricultural Year Book*, various issues.

    Pakistan Central Statistical Office, *25 Years of Pakistan in Statistics*.

    Japan Ministry of Agriculture and Forestry, *Abstract of Statistics on Agriculture, Forestry and Fisheries*, 1975.

    *Thailand Statistical Year Book*, various issues.

    *Korea Agricultural Year Book*, various issues.

    Philippines Bureau of Agricultural Economics, unpublished material.

## TABLE 10

Examples of Yield Gaps in Cultivators' Fields, India, 1977

| STATE/CROP | PRESENT AVERAGE YIELDS | PRESENT YIELDS ON BEST FIELDS[a] | INCREASE IN YIELDS POSSIBLE |
|---|---|---|---|
| | (kg/ha) | | (%) |
| **WEST BENGAL** | | | |
| Rice—Aman | 1,100 | 2,500 | 127 |
| Aus | 1,000 | 2,000 | 100 |
| Boro | 2,500 | 5,000 | 100 |
| Wheat | 2,000 | 4,000 | 100 |
| Pulses | 600 | 1,200 | 100 |
| Oilseeds | 400 | 1,000 | 150 |
| | | | |
| **MADHYA PRADESH** | | | |
| Paddy | 750 | 1,600 | 113 |
| Wheat | 730 | 2,000 | 174 |
| Pulses | 690 | 1,600 | 132 |
| Other cereals[b] | 720 | 1,800 | 150 |
| | | | |
| **RAJASTHAN** | | | |
| Wheat | 1,300 | 2,000 | 54 |
| Pulses | 500 | 1,600 | 220 |
| Other cereals[b] | 740 | 1,800 | 143 |

SOURCE: *Small Farmers and the Landless in South Asia*, World Bank Staff Working Paper no. 320 (Washington: Feb. 1979), Table 3.6, p. 33.
[a]Average on best farmers' fields; lower than obtained under research conditions.
[b]Sorghum, barley, maize, millets-barley.

## TABLE 11

### Small and Marginal Farms in South Asia

| SIZE OF OPERATED HOLDINGS | INDIA (1970-71) | | | PAKISTAN (1972) | | | BANGLADESH (1967-68) | | |
|---|---|---|---|---|---|---|---|---|---|
| | Number (millions) | Percentage of Operated: Area | Holdings | Number (millions) | Percentage of Operated: Area | Holdings | Number (millions) | Percentage of Operated: Area | Holdings |
| Less than 1 acre (near-landless) | 23.3 | 3 | 33 | 0.2 | * | 4 | 1.72 | 4 | 25 |
| Less than 2.5 acres (marginal + near-landless) | 36.0 | 9 | 51 | 0.5 | 1 | 14 | 3.9 | 21 | 57 |
| Less than 5 acres (small + marginal + near-landless) | 49.4 | 21 | 70 | 1.05 | 5 | 28 | 5.7 | 51 | 83 |

SOURCE: *Small Farmers and the Landless in South Asia*, Table 2.3, p. 14, citing: Government of Bangladesh, Bangladesh Bureau of Statistics, "Master Survey of Agriculture (Seventh Round)" (Dacca, 1972). Government of India, Ministry of Agriculture and Irrigation, "All India Report on Agricultural Census 1970-71" (New Delhi, 1975).

Government of Pakistan, Agricultural Census Organization, Ministry of Agriculture and Works, "Pakistan Census of Agriculture, 1972: All Pakistan Report" (Lahore, 1975).

*Insignificant.

TABLE 12

International Comparison of China's Level of Development in 1952

|  | USSR (1928) | JAPAN (1936) | INDIA (1950) | CHINA (1952) |
|---|---|---|---|---|
| GNP (millions) (1952 $) | 35,000 | 22,600 | 22,000 | 30,000 |
| GNP per capita (1952 $) | 240 | 325 | 60 | 50 |
| Population (millions) | 147 | 69 | 358 | 575 |
| Number of persons dependent on agriculture per acre of cultivated land | 0.20 | 1.60 | 0.60 | 1.90 |
| Paddy rice yield (tons per hectare) | 2.2 | 3.6 | 1.3 | 2.5 |

SOURCE: Alexander Eckstein, *China's Economic Development* (Ann Arbor: University of Michigan Press, 1975), p. 214, Table 7.

**TABLE 13**

Agricultural Labor Force (Midyear) Estimates, in China, 1952 to 1977

| YEAR | TOTAL POPULATION | URBAN POPULATION | RURAL POPULATION | RURAL PROPORTION OF TOTAL POPULATION | 15-64 AGE GROUP Total | 15-64 AGE GROUP Rural | RURAL LABOR FORCE | AGRICULTURAL LABOR FORCE |
|------|------------------|------------------|------------------|--------------------------------------|----------------------|----------------------|-------------------|--------------------------|
|      | (millions)       | (millions)       |                  | (percent)                            |                      | (millions)           |                   |                          |
| 1952 | 569.900 | 69.0  | 500.900 | 87.9 | 334.875 | 294.355 | 218.183 | 168.677 |
| 1953 | 582.611 | 74.6  | 508.011 | 87.2 | 340.107 | 296.558 | 219.816 | 169.940 |
| 1954 | 596.064 | 79.6  | 516.464 | 86.6 | 345.559 | 299.412 | 221.931 | 171.575 |
| 1955 | 610.201 | 82.2  | 528.001 | 86.5 | 351.185 | 303.877 | 225.241 | 174.134 |
| 1956 | 625.004 | 86.0  | 539.004 | 86.2 | 356.940 | 307.825 | 228.167 | 176.396 |
| 1957 | 640.024 | 94.3  | 545.724 | 85.3 | 362.799 | 309.345 | 229.294 | 177.267 |
| 1958 | 654.727 | 102.6 | 552.127 | 84.3 | 368.747 | 310.962 | 230.492 | 178.193 |
| 1959 | 668.930 | 109.1 | 559.830 | 83.7 | 374.679 | 313.570 | 232.425 | 179.688 |
| 1960 | 682.091 | 116.0 | 566.091 | 83.0 | 380.427 | 315.730 | 234.026 | 180.925 |
| 1961 | 693.624 | 121.2 | 572.424 | 82.5 | 386.003 | 318.555 | 236.120 | 182.544 |
| 1962 | 705.486 | 124.7 | 580.786 | 82.3 | 392.192 | 322.869 | 239.318 | 185.017 |
| 1963 | 719.301 | 128.2 | 591.101 | 82.2 | 399.458 | 328.263 | 243.316 | 188.107 |
| 1964 | 734.359 | 131.0 | 603.359 | 82.2 | 407.812 | 335.064 | 248.357 | 192.005 |
| 1965 | 750.394 | 133.8 | 616.594 | 82.2 | 417.258 | 342.858 | 254.134 | 196.471 |
| 1966 | 766.946 | 136.7 | 630.246 | 82.2 | 427.403 | 351.223 | 260.335 | 201.265 |
| 1967 | 784.107 | 142.5 | 641.517 | 81.8 | 438.222 | 358.572 | 265.782 | 205.476 |
| 1968 | 801.983 | 148.5 | 653.483 | 81.5 | 449.907 | 366.599 | 271.732 | 210.076 |
| 1969 | 820.733 | 154.7 | 666.033 | 81.1 | 462.221 | 375.097 | 278.031 | 214.946 |
| 1970 | 840.148 | 158.7 | 681.448 | 81.1 | 474.837 | 385.143 | 285.477 | 220.702 |
| 1971 | 859.927 | 162.8 | 697.127 | 81.1 | 487.741 | 395.403 | 293.082 | 226.582 |

| 1972 | 879.520 | 167.1 | 712.420 | 81.0 | 500.654 | 405.535 | 300.592 | 232.388 |
| 1973 | 898.695 | 170.7 | 727.995 | 81.0 | 513.305 | 415.807 | 308.206 | 238.274 |
| 1974 | 917.256 | 174.3 | 742.956 | 81.0 | 525.933 | 425.993 | 315.756 | 244.111 |
| 1975 | 934.626 | 177.6 | 757.026 | 81.0 | 538.642 | 436.288 | 323.387 | 250.010 |
| 1976 | 950.744 | 180.6 | 770.144 | 81.0 | 551.453 | 446.677 | 331.077 | 255.956 |
| 1977 | 965.937 | 183.5 | 782.437 | 81.0 | 564.586 | 457.315 | 338.962 | 262.052 |

SOURCE: Tang and Stone, *Food Production in the People's Republic of China*, Table 11, p. 43.

## TABLE 14
Estimates of Arable and Sown Area in China, 1949-79 (million hectares)

|  | 1949 | 1952 | 1957 | 1965 | 1970 | 1973 | 1979 |
|---|---|---|---|---|---|---|---|
| Arable area | 97.8 | 108 | 112 | 103.6 | 101.1 | 100 | 99.5 |
| Sown area | n.a. | 141 | 157 | 143 | 143.5 | 149 | 148.5 |
| Cropping index |  | 130 | 140 | 138 | 142 | 149 | 149 |

SOURCE: Ministry of Agriculture.

## TABLE 15
Area, Yield, and Production of Grains, in China, 1976-80[a]

|  | 1976 | 1977 | 1978 | 1979 | 1980 |
|---|---|---|---|---|---|
|  | (million hectares) | | | | |
| **Area** | | | | | |
| Wheat | 28.4 | 28.0 | 29.2 | 29.4 | 28.9 |
| Rice | 36.2 | 35.6 | 34.4 | 33.8 | 33.4 |
| Coarse grains | 34.0 | 33.9 | 33.5 | 33.1 | 32.7 |
| Corn | 19.2 | 19.6 | 20.0 | 20.2 | 19.9 |
| Sorghum | 4.3 | 3.8 | 3.5 | 3.2 | 3.2 |
| Millet | 4.5 | 4.5 | 4.3 | 4.2 | 4.2 |
| Barley | 4.5 | 4.5 | 4.2 | 4.0 | 3.9 |
| Oats | 1.5 | 1.5 | 1.5 | 1.5 | 1.5 |
| Others[b] | n.a. | n.a. | 23.5 | 22.7 | n.a. |
| Total[c] | n.a. | n.a. | 120.6 | 119.0 | n.a. |
|  | (tons/hectare) | | | | |
| **Yield[d]** | | | | | |
| Wheat | 1.78 | 1.46 | 1.85 | 2.13 | 1.88 |
| Rice | 3.48 | 3.61 | 3.98 | 4.25 | 4.17 |
| Coarse grains | 2.07 | 2.09 | 2.36 | 2.51 | 2.52 |
| Corn | 2.50 | 2.53 | 2.80 | 2.98 | 3.00 |
| Sorghum | 2.02 | 2.03 | 2.31 | 2.38 | 2.41 |
| Millet | 1.24 | 1.36 | 1.54 | 1.45 | 1.52 |
| Barley | 1.59 | 1.32 | 1.66 | 1.92 | 1.82 |
| Oats | 1.03 | 1.00 | 1.00 | 1.06 | 1.06 |
| Others[b] | n.a. | n.a. | 1.47 | 1.88 | n.a. |
| Total[c] | n.a. | n.a. | 2.53 | 2.79 | n.a. |
|  | (million tons) | | | | |
| **Production** | | | | | |
| Wheat | 50.5 | 41.0 | 54.0 | 62.7 | 54.2 |
| Rice | 126.0 | 128.5 | 137.0 | 143.7 | 139.3 |

*(Contd.)*

TABLE 15 *(Contd.)*

|  | 1976 | 1977 | 1978 | 1979 | 1980 |
|---|---|---|---|---|---|
|  | (million hectares) | | | | |
| Coarse grains | 70.5 | 70.7 | 79.2 | 83.0 | 82.5 |
| Corn | 48.0 | 49.5 | 55.9 | 60.0 | 59.7 |
| Sorghum | 8.7 | 7.7 | 8.1 | 7.6 | 7.7 |
| Millet | 5.6 | 6.1 | 6.6 | 6.1 | 6.4 |
| Barley | 6.7 | 5.9 | 7.0 | 7.7 | 7.1 |
| Oats | 1.5 | 1.5 | 1.5 | 1.6 | 1.6 |
| Others[b] | n.a. | 42.6 | 34.6 | 42.7 | 42.2 |
| Total[c,e] | n.a. | 282.8 | 304.8 | 332.1 | 318.2 |

SOURCE: *Agricultural Situation: People's Republic of China; Review of 1980 and Outlook for 1981*, Supplement 6 to WAS-21, USDA (Washington), Table 3, p. 20.

[a]New series based primarily on information obtained in the past two years. This series, particularly the coarse grain component, is inconsistent with the USDA historical series for years prior to 1976 (available in previous issues of this report and in various Foreign Agricultural Service Grain Circulars). No effort has been made to revise the historical series and users should be aware of the potentially misleading results obtained by combining the data presented here with the pre-1976 series.

[b]Consists of tubers (converted to a grain equivalent weight using a 5/1 conversion ratio), soybeans, pulses, and other miscellaneous grains. All of these items are included in the PRC definition of total grain. All figures in this category are calculated as a residual.

[c]PRC definition.

[d]Calculated from area and production figures.

[e]Figures for 1977-80 are official figures released by the State Statistical Bureau.

TABLE 16

Organic Fertilizer, Sources, Estimated Quantities Utilized, and Nutrient Content: China, 1952 to 1977

| YEAR | NIGHT SOIL | HOG MANURE | DRAFT ANIMAL MANURE | GREEN MANURE | OIL CAKE | COMPOST | RIVER AND POND MUD AND OTHER | TOTAL GROSS WEIGHT | NUTRIENT CONTENT |
|---|---|---|---|---|---|---|---|---|---|
| | | | | | million metric tons | | | | |
| 1952 | 185.66 | 130.13 | 422.09 | 11.2 | 4.66 | 73.71 | 114.09 | 941.5 | 10.14 |
| 1953 | 194.82 | 142.44 | 458.60 | 14.3 | 4.86 | 77.86 | 122.60 | 1,015.5 | 10.92 |
| 1954 | 205.02 | 140.33 | 491.05 | 18.3 | 4.45 | 81.12 | 131.74 | 1,072.0 | 11.41 |
| 1955 | 215.85 | 130.67 | 511.35 | 23.5 | 4.47 | 82.25 | 141.56 | 1,109.7 | 11.72 |
| 1956 | 227.41 | 152.88 | 516.82 | 30.1 | 5.01 | 81.00 | 152.12 | 1,165.3 | 12.37 |
| 1957 | 237.15 | 184.00 | 514.11 | 38.5 | 4.92 | 78.56 | 152.12 | 1,209.4 | 13.03 |
| 1958 | 241.75 | 256.00 | 523.97 | 49.2 | 5.14 | 80.06 | 152.12 | 1,307.9 | 14.23 |
| 1959 | 245.75 | 217.76 | 503.84 | 40.6 | 5.63 | 76.99 | 152.12 | 1,242.7 | 13.48 |
| 1960 | 249.64 | 168.85 | 490.19 | 39.4 | 4.02 | 74.90 | 152.12 | 1,179.4 | 12.50 |
| 1961 | 253.21 | 162.76 | 452.21 | 39.4 | 3.87 | 69.10 | 152.12 | 1,132.7 | 12.00 |
| 1962 | 257.25 | 184.74 | 436.56 | 44.3 | 3.77 | 66.71 | 152.12 | 1,145.5 | 12.20 |
| 1963 | 272.87 | 226.79 | 490.72 | 59.9 | 3.45 | 72.27 | 152.12 | 1,278.1 | 13.67 |
| 1964 | 292.59 | 258.04 | 555.19 | 81.0 | 3.40 | 78.01 | 152.12 | 1,420.4 | 15.23 |
| 1965 | 310.92 | 301.79 | 631.94 | 85.7 | 3.35 | 85.83 | 152.12 | 1,571.6 | 16.95 |
| 1966 | 317.79 | 329.52 | 640.33 | 90.6 | 3.33 | 86.97 | 152.12 | 1,620.6 | 17.56 |
| 1967 | 324.31 | 357.62 | 598.85 | 95.8 | 3.41 | 81.34 | 152.12 | 1,613.4 | 17.60 |
| 1968 | 331.19 | 356.91 | 580.69 | 101.3 | 3.18 | 78.87 | 152.12 | 1,604.3 | 17.48 |
| 1969 | 338.39 | 343.45 | 602.91 | 107.1 | 3.04 | 81.89 | 152.12 | 1,629.0 | 17.67 |
| 1970 | 346.33 | 378.64 | 608.47 | 113.3 | 3.38 | 82.65 | 152.12 | 1,684.9 | 18.40 |
| 1971 | 354.41 | 415.66 | 638.46 | 119.8 | 3.87 | 86.72 | 152.12 | 1,771.0 | 19.47 |
| 1972 | 362.36 | 467.78 | 658.63 | 126.7 | 4.26 | 89.46 | 152.12 | 1,861.4 | 20.63 |
| 1973 | 370.27 | 420.20 | 678.73 | 134.0 | 4.90 | 92.19 | 152.12 | 1,852.6 | 20.42 |

| 1974 | 377.91 | 455.04 | 698.82 | 141.7 | 4.65 | 94.92 | 152.12 | 1,925.2 | 21.25 |
| 1975 | 385.06 | 482.79 | 719.47 | 149.9 | 4.90 | 97.72 | 152.12 | 1,992.0 | 22.08 |
| 1976 | 391.71 | 489.78 | 740.75 | 158.5 | 4.41 | 100.62 | 152.12 | 2,037.9 | 22.52 |
| 1977 | 397.97 | 492.30 | 762.72 | 167.6 | 4.66 | 103.60 | 152.12 | 2,081.0 | 22.99 |

SOURCE: Tang and Stone, *Food Production in People's Republic of China*, Table 17, p. 61.

## TABLE 17

The Composition of Total Number of Agricultural Shūraku (Hamlet) by Method of Meeting the Labor Demand for Joint Operations (all Japan except Hokkaido, 1970)

| | MAINTENANCE AND REPAIRING OF ORDINARY ROADS | MAINTENANCE AND REPAIRING OF AGRICULTURAL ROADS | WEEDING AND MUD-REMOVING OF FARM DITCHES FOR IRRIGATION AND DRAINAGE |
| --- | --- | --- | --- |
| | % of Total Number of Agricultural Shūraku (135,206) | | |
| 1. Undertaken by joint operations of Shūraku | 73.6 | 74.0 | 63.8 |
| a. All the families of Shūraku should provide labor | 53.1 | 52.0 | 43.6 |
| b. Those which do not provide labor should pay money for hiring labor | 17.7 | 17.3 | 14.4 |
| c. Those persons who provide labor are paid wages | 1.9 | 2.7 | 14.4 |
| d. Others | 0.9 | 1.9 | 2.7 |
| 2. Undertaken by Shūraku by hiring labor | 0.3 | 0.2 | 0.3 |
| 3. Not undertaken by Shūraku | 26.1 | 25.8 | 36.0 |

SOURCE: Ministry of Agriculture and Forestry, *1970-nen Sekai Nōringyō Sensasu—Nōgyō Shūraku Chōsa Hōkokusho* (1970 World Agricultural and Forestry Census—Report on the Survey of Agricultural Shūraku) (Tokyo, 1972). Cited in Shigeru Ishikawa, *Employment Expansion in Asian Agriculture*, Table 7, p. 54.

TABLE 18

Changes in the Principal Methods of Production behind the Changes in Labor Inputs in Rice Production in Pre-War Saga Plain

| MAN-DAYS/HA | END OF TOKUGAWA ERA 350-400 | MID-MEIJI 280-300 | EARLY TAISHO ca 280 | LATE TAISHO ca 200 | ABOUT SHOWA 10 ca 140 |
|---|---|---|---|---|---|
| Amount of decrease of man-days from the previous period | | (−) 70-100 | Slight decrease | (−) ca 80 | (−) ca 60 |
| **MAIN OPERATIONS** | | | | | |
| Preparatory tillage | horse tillage *plus* plough with long bed | | | improved plough with short bed | a further improved type (simplified the task of controlling horse tillage |
| Transplanting | irregular-lines | straight-lines | | | |
| Weeding | Kari-tsume (hand tool for weeding) | Tauchi-guruma (weeder) | | | |
| Fertilizing | mud collection from creeks | gradual introduction of purchased fertilizer | | chemical fertilizer | |
| Irrigation | water-basket treadle wheel | | | electric pumps | |
| Threshing | Senba (comb-type threshing soil) | | | pedal threshers | mechanical threshers |
| Hulling | hand mortar | | | rice hullers | |

SOURCE: The works of Toshihiko Isobe, Tatsuo Yamada and Ryoichiro Ota, and Isao Kamagata. See S. Ishikawa "Azia Nōson no Kōyo Mondai" (Employment Problems in the Rural Regions of Asia), *Keizai Kenkyu* (April 1979), Table. Cited in Shigeru Ishikawa, *Employment Expansion in Asian Agriculture*, Table 1, p. 48.

# APPENDIX—AFRICAN STATISTICS

## TABLE 19

Cropland per Person in Sub-Saharan Africa by Region and Country[a]

| REGION AND COUNTRY | NUMBER OF PERSONS PER KM² CROPLAND | REGION AND COUNTRY | NUMBER OF PERSONS PER KM² CROPLAND |
|---|---|---|---|
| SAHEL: | 69.6 | Congo | 212.3 |
| Cape Verde | 750.0 | Gabon | 125.1 |
| Chad | 60.0 | Zaire | 417.5 |
| Gambia | 200.0 | | |
| Mali | 61.1 | EAST AFRICA: | 275.5 |
| Mauritania | 682.7 | Burundi | 311.6 |
| Niger | 31.6 | Ethiopia | 213.6 |
| Senegal | 192.8 | Kenya | 623.9 |
| Upper Volta | 112.5 | Rwanda | 464.7 |
| | | Somalia | 312.5 |
| WEST AFRICA: | 182.9 | Sudan | 259.6 |
| Benin | 110.1 | Tanzania | 321.1 |
| Cameroon | 90.3 | Uganda | 217.7 |
| Ghana | 386.9 | | |
| Guinea | 111.3 | SOUTHERN AFRICA: | 208.5 |
| Guinea-Bissau | 315.8 | Botswana | 536.8 |
| Ivory Coast | 56.2 | Lesotho | 338.0 |
| Liberia | 483.8 | Madagascar | 276.5 |
| Nigeria | 277.5 | Malawi | 226.5 |
| Sierra Leone | 76.4 | Mozambique | 314.1 |
| Togo | 104.1 | Namibia | 152.8 |
| | | Swaziland | 297.0 |
| CENTRAL AFRICA: | 241.7 | Zambia | 107.3 |
| Angola | 364.4 | Zimbabwe | 271.3 |
| Cent. African Republic | 31.7 | | |

SOURCE: Economic, Statistics, and Cooperatives Service: Food and Agriculture Organization, *Production Yearbook* (1978); USDA, *Food Problems and Prospects in Sub-Saharan Africa* (1981).

[a]Cropland refers to land defined by the FAO as arable land and land under permanent crops. It includes land under temporary crops, temporary meadows for mowing and pasture, land under market and kitchen gardens, land temporarily fallow or idle, and land cultivated with crops that occupy the land for long periods and need not be replanted, such as rubber, cocoa, and coffee. Definitions used by reporting countries vary, however, so that classification of different kinds of land may be inconsistent.

TABLE 20

Populations in Sub-Saharan Africa and South Asia

|  | TOTAL POPULATION (mid-1979) |
|---|---|
|  | (millions) |
| Sub-Saharan Africa | 343.9 |
| Low-income | 187.1 |
| Nigeria | 82.6 |
| Other middle-income | 74.2 |
| South Asia[a] | 890.5 |

SOURCE: *World Bank Reports.*
[a]Bhutan, Bangladesh, Nepal, Burma, India, Sri Lanka, and Pakistan.

TABLE 21

Area, Yield, and Production: Average Annual Growth Rates, Sub-Saharan Africa, 1962/64-1972/74

| COMMODITY AND REGION | PRODUCTION | YIELD | AREA |
|---|---|---|---|
|  | (percent) | | |
| Cereal: | | | |
| Sahel | −1.6 | −1.3 | −0.3 |
| West Africa | 0.7 | −0.6 | 1.3 |
| Central Africa | 3.2 | −1.6 | 4.7 |
| East and Southern Africa | 2.6 | 1.5 | 1.1 |
| Roots and tubers: | | | |
| Sahel | −0.7 | −1.2 | 0.5 |
| West Africa | 1.8 | 0.4 | 1.4 |
| Central Africa | 1.4 | −1.7 | 3.1 |
| East and Southern Africa | 1.7 | 3.3 | −1.5 |
| Pulses: | | | |
| Sahel | 1.6 | −1.9 | 3.5 |
| West Africa | 2.0 | −2.6 | 4.8 |
| Central Africa | 1.9 | −2.0 | 4.0 |
| East and Southern Africa | 1.9 | −0.2 | 2.1 |

SOURCE: Food and Agriculture Organization, *Regional Food Plan for Africa* (July 1978).

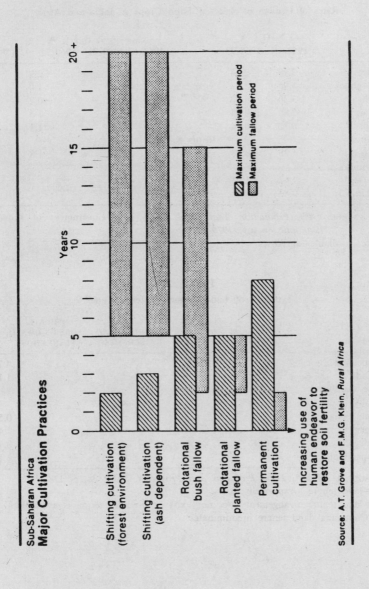

Sub-Saharan Africa
**Major Cultivation Practices**

Source: A.T. Grove and F.M.G. Klein, *Rural Africa*

TABLE 22

Rates of Growth of Selected Food Crops in India and Africa

| | INDIA | | | AFRICA | |
| | 1952-65 | 1967-79 | | 1961/63-1969/71 | 1969/71-1977/79 |
|---|---|---|---|---|---|
| Rice | 3.18 | 2.64 | | 4.0 | 2.9 |
| Sorghum | 1.96 | 2.07 | + millet | 0.9 | 1.0 |
| Wheat | 3.30 | 6.02 | | 3.8 | −0.2 |
| Maize | 2.80 | −0.04 | | 5.2 | 1.3 |
| Pulses | 0.72 | 0.54 | | 3.3 | 1.1 |
| Bajra* | | | Roots & | | |
| (pearl | 1.38 | 0.28 | tubers | 2.0 | 1.8 |
| millet) | | | | | |
| Barley* | −1.62 | −1.95 | | | |
| Gram* | 0.83 | 0.66 | | | |

SOURCE: FAO, *Production Year Book* tapes; and Government of India, *Economic Survey, 1979-80*.

*Not available for Africa.

TABLE 23

Typology of Land Tenure Patterns, Sahel[a]

| COUNTRY | INDIVIDUAL TITLE | STATE FARM | CONTROLLED SCHEMES | PRIVATE, FOREIGN-OWNED PLANTATIONS |
|---|---|---|---|---|
| Cape Verde | − | × | − | − |
| Chad | − | − | − | − |
| The Gambia | − | − | − | − |
| Mali | − | − | × | − |
| Mauritania | − | − | − | − |
| Niger | − | − | − | − |
| Senegal | × | − | × | − |
| Upper Volta | − | − | × | − |

SOURCE: USDA, *Food Problems and Prospects*, Table 38, p. 138.

×= land tenure arrangement exists.

−= land tenure arrangement does not exist or no information is available.

[a]Communal land tenure predominates.

## TABLE 24
### Typology of Land Tenure Patterns, West Africa[a]

| COUNTRY | INDIVIDUAL TITLE | STATE FARM | CONTROLLED SCHEMES | PRIVATE, FOREIGN-OWNED PLANTATIONS |
|---|---|---|---|---|
| Benin | X | X | — | — |
| Cameroon | X | X | — | — |
| Ghana | X | X | — | — |
| Guinea | X | — | — | — |
| Guinea-Bissau | X | X | — | — |
| Ivory Coast | X | X | — | — |
| Liberia | X | — | — | X |
| Nigeria | X | — | — | — |
| Sierra Leone | X | — | — | — |
| Togo | X | X | — | — |

SOURCE: USDA, *Food Problems and Prospects*, Table 43, p. 154.
X = land tenure arrangement exists.
— = land tenure arrangement does not exist or no information is available.
[a] Communal land tenure predominates.

## TABLE 25
### Typology of Land Tenure Patterns, Central Africa[a]

| COUNTRY | INDIVIDUAL TITLE | STATE FARM | CONTROLLED SCHEMES | PRIVATE, FOREIGN-OWNED PLANTATIONS |
|---|---|---|---|---|
| Angola | — | X | X | — |
| Central African Republic | X | — | — | X |
| Congo | X | X | — | — |
| Gabon | X | — | — | X |
| Zaire | X | — | X | X |

SOURCE: USDA, *Food Problems and Prospects*, Table 48, p. 172.
X = land tenure arrangement exists.
— = land tenure arrangement does not exist or no information is available.
[a] Communal land tenure predominates.

TABLE 26
Typology of Land Tenure Patterns, East Africa[a]

| COUNTRY | INDIVIDUAL TITLE | STATE FARM | CONTROLLED SCHEMES | PRIVATE. FOREIGN-OWNED PLANTATIONS |
|---|---|---|---|---|
| Burundi | − | ✕ | ✕ | − |
| Ethiopia[b] | − | ✕ | − | − |
| Kenya | ✕ | − | − | ✕ |
| Rwanda | − | − | ✕ | − |
| Somalia[b] | ✕ | ✕ | ✕ | − |
| Sudan | ✕ | ✕ | ✕ | ✕ |
| Tanzania[b] | ✕ | ✕ | − | ✕ |
| Uganda | ✕ | − | − | ✕ |

SOURCE: USDA, *Food Problems and Prospects*, Table 53, p. 180.
✕ = land tenure arrangement exists.
− = land tenure arrangement does not exist or no information is available.
[a]Communal land tenure predominates.
[b]Some collectivized holdings exist.

TABLE 27
Typology of Land Tenure Patterns, Southern Africa[a]

| COUNTRY | INDIVIDUAL TITLE | STATE FARM | CONTROLLED SCHEMES | PRIVATE, FOREIGN-OWNED PLANTATIONS |
|---|---|---|---|---|
| Botswana | ✕ | − | − | ✕ |
| Lesotho | − | − | − | − |
| Madagascar | ✕ | ✕ | ✕ | ✕ |
| Malawi | ✕ | − | − | ✕ |
| Mozambique | ✕ | ✕ | − | ✕ |
| Namibia | ✕ | − | − | ✕ |
| Swaziland | ✕ | − | − | ✕ |
| Zambia | ✕ | ✕ | − | ✕ |
| Zimbabwe | ✕ | − | − | ✕ |

SOURCE: USDA, *Food Problems and Prospects*, Table 58. p. 200.
✕ = land tenure arrangement exists.
− = land tenure arrangement does not exist or no information is available.
[a]Communal land tenure predominates.

TABLE 28

Modern Input Use, Africa, Asia, and South America, 1977

| AREA | PERCENTAGE OF IRRIGATED LAND | TRACTORS PER 10,000 HECTARES | FERTILIZER USED PER HECTARE |
|------|------|------|------|
| | (percent) | (number) | (kilograms) |
| Africa | 1.8 | 7 | 4.4 |
| Asia | 28.0 | 45 | 45.4 |
| South America | 6.1 | 57 | 38.8 |

SOURCE: Food and Agriculture Organization, *Production Year Book* (1978), and *Fertilizer Year Book* (1978).

TABLE 29

Estimated Investment Requirements for Closing the Food Gap, 24 Selected Countries, Sub-Saharan Africa, 1975-90

| TYPE OF INVESTMENT | AMOUNT OF INVESTMENT REQUIRED | |
|------|------|------|
| | IFPRI | FAO[a] |
| | (000 1975 US $) | |
| Irrigation infrastructure | 3,132,100 | 4,784,000 |
| Training personnel for irrigation | 20,008 | NI |
| Settlement of rain-fed land | 830,000 | 1,266,000 |
| Road construction | 859,000 | NI |
| Electrification | 4,311,000[b] | NI |
| Fertilizer manufacture[c]. | 506,952 | NI[d] |
| Improved seeds | 18,960 | NI[e] |
| Mechanization | 702,724[a] | 5,153,000 |
| Pesticide supply | 96,655 | NI |
| Storage improvement | 1,065,666 | NI |
| Research and extension | 1,126,400[a] | NI |
| Livestock development | NI | 3,831,000 |
| Total | 12,699,465 | 15,034,000 |

SOURCE: As cited in USDA, *Food Problems and Prospects*, Table 34, p. 118: Peter Oram, Juan Zapata, George Alibaruho, and Shyamal Roy, *Investment and Input Requirements for Accelerating Food Production in Low-Income Countries by 1990* (Washington: International Food Policy Research Institute, December 1979), various pages. Food and Agriculture Organization, *Regional Food Plan for Africa* (Rome: July 1978), Table E-1.

NI = not included.

[a]Less Sudan.

[b]IFPRI allocated $6 billion in U.S. currency to Africa for settlement of rain-fed land, road construction, and electrification (p. 85).

[c]Calculated at $240 in U.S. currency per metric ton.

[d]Cost of fertilizer inputs estimated at $2,328,000,000 in U.S. currency annually by 1990.

[e]Cost of seed inputs estimated at $323,000,000 in U.S. currency annually by 1990.

# INDEX